EVERYWHEN

God, Symmetry, and Time

Thomas P. Sheahen, Ph.D.

En Route Books and Media, LLC

St. Louis, MO

ENROUTE
Make the time

En Route Books and Media, LLC
5705 Rhodes Avenue
St. Louis, MO 63109

Cover credit: NASA Hubble Space Telescope

ISBN-13: 978-1-956715-10-1

Library of Congress Control Number: 2021950968

Table of Contents

Appendices

Acknowledgments

Most important of all, I want to thank my daughter, Laura Sheahen, for her exceptional talent and initiative in making this a real book. Laura perceived the possibility and poured enormous effort over a two-year period into blending disparate essays into a coherent framework. Both her insights into methods of communicating and her editing skills have been invaluable. There would be no "Everywhen" book without her.

Fr. Robert A. Brungs, S.J. (1931-2006) was very helpful across several decades, encouraging me in pursuing original and unorthodox lines of thought. The organization he founded, the *Institute for Theological Encounter with Science & Technology* (ITEST) is dedicated to showing that faith and science are not only compatible but complementary as pathways toward knowledge of God.

Sr. Marianne Postiglione, R.S.M., the associate director of ITEST and former editor of the quarterly *ITEST Bulletin*, has also been entirely supportive of my writing efforts over the dozen years of my tenure as director of ITEST. Marianne has been a full partner in the leadership of ITEST.

Several friends and professional colleagues have read drafts of the book and identified significant points needing revision and clarification. Prominent among them is Fr. Bob

Spitzer, S.J., whose outstanding advocacy within the Church of the viewpoint that faith and science are compatible was what inspired me to get going and actually write *Everywhen*. Additionally, I'm indebted to John Shaw and Cal Beisner for their insights and encouragement.

About the Cover Photo

The Hubble Space Telescope photographed the *Veil Nebula*, the remnants of a supernova about 2,100 light years away. The dramatic colors are indicative of the various ions radiating (Oxygen, Nitrogen, etc.). Trying to express our very imperfect understanding of God, St. Paul once said "For now we see only a reflection as in a mirror; then we shall see face to face" (1 Cor. 13:12). Another form of impaired vision is through a veil, which gives a cloudy, blurry image. The many stars and galaxies behind the *Veil Nebula* remind us of how faulty our vision of God is.

Preface

Faith and Science: Are They Really Opposed?

Are science and religion opposed? Is it possible to be a religious believer – and also believe in science?

Atheists get a lot of attention because they boost the notion that the two are in conflict. And some (not all!) scientists assert that the universe created itself, with no involvement by God; hence, there is no need for God. That's a faulty exaltation of science. Rather, it's just as important to know what science *can't* cover as it is to know what science *does* cover.

In this book, I present examples of new ways of thinking. I try to overcome barriers that have caused people to think in restricted ways. My hope is that you will react by escalating your own thinking to a higher plane, seeing reality, nature, life, and spirituality differently. If you're motivated to imagine new realms and explore them, then this book fulfills its purpose.

My guiding principle starts with the words of the early Christian philosopher St. Augustine, who said: "The book of nature and the book of Scripture were both written by the same Author, and they will not be in conflict when properly read and interpreted." That's the cornerstone of my approach.

As one example, St. Augustine figured out that God *created* space and time *together*, and that was "the beginning." He recognized that our coordinate system is not just there, but was *created* by God, and there could be no "before" or "after" in the absence of time.

Sadly, Augustine's wisdom was forgotten over centuries, and science developed with the presumption that *time* is absolute, that it's always been there, that it just *is*. "Time" somehow appeared more important than God. Big mistake.

To correct that error, I explain how modern science now understands the fundamental importance of symmetry, including the interrelatedness of space and time. Prior to the 20th century, nobody recognized that symmetry principles (which God also created) lie hidden at the very core of the laws of physics … and, in turn, of chemistry and the life sciences.

As a physicist, the appreciation I have for the symmetry and beauty of the laws of physics points toward the magnificent power of God. Many other scientists draw upon their own disciplines (biology, medicine, engineering, …) to appreciate what God has accomplished. Each individual is able to pursue his/her own distinctive pathway toward that goal of reconciling faith and science.

Nevertheless, certain things that "everybody knows…" get in the way. To this day, most people still conceive of time as linear – going from the past to the present to the future. Plenty of scientists formulate their conception of God within that restricted framework and wind up imagining a god that is subordinate to time – quite inferior to the God who is the *creator* of time. After that, they invent reasons to disbelieve in such a limited god.

That kind of "either/or" limited thinking needs to be overcome by climbing upward to a new plateau of thinking and posing questions, where "both/and" answers make perfectly good sense. The stairs to that higher level begin with very simply trusting that God is smarter and more original than even our best theories.

In thinking up "science," God made sure it is a parallel path alongside religious faith, where both can lead mankind toward God. Properly reading the book of nature requires a form of faith as well, and God has planted plenty of surprises along the way. Having confidence that God is going to provide "both/and" answers makes the path less steep.

The grandeur of God's creation greatly exceeds our language and thought processes, which act as impediments. To explore new territory, I borrow from geometry the concept of higher dimensions, a device to aid perception. Expanding our horizon into a higher-dimensional domain, we concede that God fully comprehends all additional dimensions to which humans have access.

From that new plateau, the view looks different, old obstacles have fallen behind, and new possibilities come into sight (although perhaps clouded in fog). But we can be sure that there are still higher levels that challenge others to tackle.

If you have felt torn, thinking you must choose between believing in God OR believing in science, don't be discouraged. You can be confident that faith and science are actually partners on the path toward knowledge.

Thomas P. Sheahen

November 2021

Part 1

Bridging Science and Faith

From the outset, Christianity always held that God made a world which makes sense. Roman gods warring against each other did not. Two millennia later, that cornerstone belief has brought us to a very advanced state of science and engineering, the influence of which penetrates everyone's life. Today, there are those who say science is all we need, and faith in God is obsolete, unnecessary baggage. That's the very definition of *hubris*. To avoid that mistake, we must remember that God invented science.

In these first chapters, we take a close look at science and discover how intertwined it is with faith. The two really are inseparable; neither gets very far without the other. At its foundational level – the laws of physics – science is based on *symmetry principles* created by God. The symmetry that God created between space and time is real but almost invisible to mankind. That deficiency presents a major obstacle to understanding God and His creation. The route to overcoming that deficiency requires a departure from conventional thinking.

That departure – taking an upward step to a higher level of thinking – allows us to look beyond the confines of space and time, to appreciate mankind as a higher-level being. Using the insight of higher dimensionality, we can glimpse the magnificent way that God has enabled His creation (us) to interact with Him.

Chapter 1

God Is Everywhere and *Everywhen*

Everyone naturally has questions about God and would like to improve the understanding each has of God. Certain characteristics of God are widely acknowledged. God, for instance, must be all-powerful (*omnipotent*) and all-knowing (*omniscient*). Being spiritual, God is both *transcendent* over his creation and *immanent* within it. To support these capabilities, God must also be *omnipresent*, always aware.

A major obstacle, however, stands in the way. We humans have a very limited perception of *time*, which affects our understanding of God's comprehension of time – the quality of *Omnipresence*. Our experience of time (quite inferior to God's) is a blockade to our appreciating and trusting God. Once we realize that God is not constrained by time, some long-standing controversies and religious paradoxes dissolve. It also becomes easier to reconcile modern science with the Bible.

Omnipresence

When asked what "omnipresence" means, most people would reply "God is everywhere." But it also means that God

is every*when*, that is, *present* to all *time*. That is totally beyond human comprehension. The prefix "omni" is attached
to a term that we *do* comprehend, so we seldom notice the
enormity of the gap.

As human understanding of God has improved, human
thought, language, and culture have caused some totally
human limitations to creep in, unnoticed. Contemporary
perception holds that God shares certain properties with
mankind; moreover, God is presumed bound by certain
human constraints. It "makes sense" because we're unable to
think any other way. But that's where errors come from.

To say *Omnipresence* and mean "God is everywhere" is
comfortable to most people. But to say "God is *present* to all
time" is quite another matter. People may nod in agreement
with a theological statement like that but can't really inter-
nalize it. The word *everywhen* is absent from conventional
languages. The idea of God's perceiving *all* time – ancient,
now, future – in some unified way called "present" is
incomprehensible to nearly everyone. But the significant
point is this: that great difficulty is due to a *human* limitation.

Unfortunately, there has been a near-universal failure to
recognize that deficiency, and instead people have assumed
that God must perceive time in the same way we do –
including that He must deal with time sequentially. That is a
major error, which causes an impression of a weakness on

God's part. In that viewpoint, God exists *within* time, and God is *subordinate* to time.

People have forgotten that God is the *Creator* of time. "Time just IS" is the common perception, and *Time Marches On* is a standard slogan. Ordinary experience seems to indicate very firmly that time is absolutely immutable. To appreciate that God transcends time requires the mental effort of an excursion beyond customary thinking.

Fortunately, modern science gives a boost in that direction, by presenting facts quite differently from that historical human perception. We have reached the point where the *Hubble Space Telescope* shows us galaxies at enormous distances, but also very far back in time. If we, with limited equipment and limited mental capabilities, can appreciate that expanse, can we doubt that God encompasses it all, and is simply *present* to everything?

Overcoming False Constraints

There are a number of consequences for humans if we "allow" God to have the attribute of being *present* to all time. Here is a very essential guideline: *Never underestimate God.* Do not assume that God is subject to the same limitations as people.

Too often, questions about God are posed in a framework that expects "either/or" answers, and irresolvable disagreements ensue among conflicting groups. But the unexpected "right answers" more likely have the form "both/and" at a higher level of thinking. The better ap-

> **A very essential guideline: Never Underestimate God.**

proach is to locate that higher level and step up to it.

The realm of God's creation goes far beyond our ability to think and express ideas; language itself poses impediments. Recognizing such limitations is the first step toward overcoming them. Science can help achieve that goal. Of everything God created, *science* is an exceptionally magnificent gift, which usually goes unappreciated.

Today, we recognize science as more than just a collection of facts; it is an intelligent, organized way of understanding and dealing with everything around us. But it's very easy to forget that God gave coherence and intelligibility to all creation; that is, God *thought up* science. Such forgetting leads a lot of people to think that science has nothing to do with God, or with the way we relate to God.

Fortunately, God invites us to travel along a superior pathway to knowledge, which proceeds by striving to find the *unity* between science and religious faith.

Pope St. John Paul II expressed the relationship between religion and science by saying: "Science can purify religion from error and superstition. Religion can purify science from idolatry and false absolutes. Each can draw the other into a wider world, a world in which both can flourish."

Scientists are quick to criticize superstitions, but they seldom detect the false absolutes they have imposed on their own understanding. The notion that "if it can't be explained by science, it doesn't exist" is typical of how some scientists deliberately avoid being drawn into a wider world. By placing a narrow boundary around their turf, and then insisting that everyone else conform to their style of thinking, they don't even see the false absolute they have built.

Time is the most insidious and unrecognized false absolute, from which everyone suffers. That obstacle is the topic of chapter 4.

Stepping Upward

Once we acknowledge that human thought is severely limited and gets in the way of relating to God, it becomes possible to take an upward step in thinking, to explore new territory.

For thousands of years, humans have tried various ways to express the notion that we are much *more* than our senses perceive. Plato's "cave" analogy[1] is one early example that borrowed from mathematics: Plato said reality is multi-dimensional, but all we experience is the downward projection to fewer dimensions (the shadow on the cave wall). In the chapters ahead, the geometric concept of higher dimensions

---

[1] Plato, *The Republic,* Book 7.

will be employed as a mechanism to aid perception of true humanity.

Shifting our thinking into a higher-dimensional domain, we immediately acknowledge that God has a total grasp of all such additional dimensions (because He's a better mathematician). That line of thinking might yield insights into higher properties of humans. Eventually, however, the problem of encountering something entirely beyond experience occurs. In that case, the applicable word is *trans-intelligible* (which is quite different from unintelligible). A statement can be true and yet beyond the reach of a finite mind. The word *mystery* is commonly used in spiritual and theological writings to convey the same notion.

This has many important applications, in surprising ways. For example, nearly everyone tries to formulate a personal concept of life after death. It's an "eternal question." Poets, writers, and mystics have offered a wide spectrum of images over the millennia, and all agree the images are inaccurate. The idea of dimensions that are separate and distinct from time, not bound by time, offers a new avenue for original exploration. Any new images will still be inaccurate, but the task is not hopeless.

Resolving Old Problems

Perhaps the first to think of space and time jointly was St. Augustine, who wrote circa 400, long before any organized science began. Augustine said that God created space and time together,[2] and that was the beginning of creation. The ancient Greeks had just taken the *coordinate system* for granted, but Augustine pointed out that it too is a creation of God. This may well be the single most underrated achievement in the entire realm of religion and science.

Augustine also solved the time-dependent riddle: "What was God doing before the creation?" by noting that the word "before" has no meaning whatsoever until "after" the creation of space and time. The key point to grasp here is that Augustine placed space and time in a position *subordinate* to God.

Sadly, Augustine's insight was forgotten, and the science of physics got under way in the 17th century with the faulty perception that the coordinates (space and time) are immutable, not a creation of God.

Outside the realm of physics, major difficulties result from imposing the limited human perception of time when constructing an image of God. Here are three such topics that will be covered later in this book:

1. The entire argument about *predestination* is rooted in the 'either/or" position that God supposedly must take with regard to events happening sequentially in

---

[2] St. Augustine, *The City of God*, Book XI, section 6 (New York: Random House, 1950).

time. Because of the human way of considering time, it is mind-boggling to imagine that God could know the future without forcing the past. To rise above an "either/or" position, adopting a "both/and" position, requires stepping up to a more advanced level of thinking.

2. *Process Theology* imagines that God changes and develops with the passage of time. This idea began late in the 19th century when classical mechanics (containing absolute time) was at its zenith and determinism was believed to be built into the laws of nature. To get away from static determinism, it seemed plausible to have God change over time, just as human beings do. After a century of progress in modern physics, determinism has been swept away, but the supremacy of absolute time lingers on. It is certainly valid to say that mankind's *perception* of God develops over time, but that doesn't mean that God changes over time.

3. *Creationism:* Atheists point out that evolution has taken a really long time, which implies that any god subject to time must be weak. Sadly, too many religious people accept their premise and hence face a dilemma. Creationists, who reject the age of the universe, have fallen into the trap of imagining a contradiction between *God's having* a purpose and *God's utilizing* extremely long times to make it come

true. It's only a contradiction if God is subordinate to time. This is treated more in Chapter 8.

The significant point to note about each of these three issues is that they came to prominence by failing to notice a human limitation being imposed on God.

God's Interactions with Humanity

It is important not to claim too much. No scientific theory is ever going to address satisfactorily a great number of theological questions. For instance:

- Can God, who is the creator of time and transcends time, enter into a relationship with mankind and yet preserve His transcendence to time? The entire Bible replies "yes."
- Can God choose to enter into humanity with all its limitations, including becoming subordinate to time? The Christian answer is "yes." Through the *incarnation*, God did precisely that; Jesus Christ lived at a particular time and place. Christianity would say further that, from the very start, God oriented His creation with this in mind.
- Here is yet another question: What does *resurrection* mean for a standard human being? Does it have to be linked to time?

These questions remain in the realm of mystery – they are beyond human comprehension, that is, *transintelligible*.

> **Properly understood, omnipresence applies to both space and time. God is everywhere. God is also Everywhen.**

Everyone will agree that we have only limited images of God. Still… we try to make progress by associating certain terms with the various attributes of God. Then we strive to understand what those terms mean. We don't always succeed and must concede that we are *not* going to understand how God comprehends time, no matter how hard we try.

Properly understood, *omnipresence* applies to both space and time. God is everywhere. God is also *everywhen*.

Everywhen certainly exists at the level of the mind of God, but we can't access it because we are incapable of matching God's ability to comprehend all reality – past, present, future – "simultaneously." We employ the clumsy term "all at once" as if to imply that God has merely compressed time, which humans still perceive as sequential.

Being stuck in that "sequential" mode is quite a handicap. We need to readjust our thinking and recognize that a human limitation has seriously curtailed our understanding of God. The best we can do is recognize that, from God's viewpoint, *everywhen* is perfectly normal.

For us, that's progress; but there is still a very long way to go. Great scientists such as Newton and Einstein (see next page) have commented on how primitive our science is. Humility like that is essential for future progress.

Nothing said here even begins to address the central mysteries of theology, involving God's *transcendence* and *immanence*. In future times, others (free of past constraints) will think and understand at a new, higher level. That higher level will enable them to achieve an enhanced relationship with God.

Sir Isaac Newton and Albert Einstein

SIR ISAAC NEWTON

ALBERT EINSTEIN

"I do not know what I may appear to the world, but to myself I seem to have been only like a boy playing on the sea-shore, and diverting myself in now and then finding a smoother pebble or a prettier shell than ordinary, whilst the great ocean of truth lay all undiscovered before me."

"We are in the position of a little child entering a huge library filled with books in many languages. The child knows someone must have written those books. It does not know how. It does not understand the languages in which they are written. The child dimly suspects a mysterious order in the arrangement of the books but doesn't know what it is. That, it seems to me, is the attitude of even the most intelligent human being toward God. We see the universe marvelously arranged and obeying certain laws but only dimly understand these laws."

Chapter 2

Symmetry, Time, and Space

In this chapter, we examine the very basic and elegant principles that underlie science, which God used to create a universe that makes sense and is orderly. From those principles come the "laws of physics." Then in turn, the entire universe and everything we see around us is a secondary creation that proceeds from His initial creation. The familiar bromide "you can't make this up" testifies to the stunning brilliance of God's wisdom.

Symmetry and Conservation Laws

One example or symbol of perfection has always been the circle. It recurs in artwork from time immemorial. The association of beauty with *symmetry* is universal in human imagination. For thousands of years, astronomers took data on the positions of heavenly bodies. They were trying to discern regularity and order, using circular motion as the cornerstone of their thinking.

Somewhere over a millennium ago, the idea emerged that nature made sense, that it is subject to rational thought, that it can be understood through diligent investigation.

Thus, there began what became known as *the Scientific Method*. Several centuries later, people doing organized investigations of nature conducted experiments and examined the data in search of regularities, patterns, or rules. In the everyday world of simple objects Galileo Galilei demonstrated that mathematics can be a useful tool to describe movement.[1] He expressed the laws of motion as mathematical formulas.[2]

Later, Isaac Newton originated *classical mechanics,*[3] a mathematical description of nature that expressed the relations between forces and energy.[4] In particular, there were three major *Conservation Laws*: The Conservation of Energy, the Conservation of Momentum, and the Conservation of Angular Momentum. Experiments always confirmed the validity of the conservation laws, and they became accepted so well that no theory alleging a violation would be considered at all.

---

[1] Antonino Zichichi, *Galilei Divine Man* (Rome: Il Cigno GG Edizioni, 2018).

[2] Many books cover this history. One fairly easy read is: George Gamow, *Matter, Earth & Sky* (Upper Saddle River, N.J.: Prentice Hall, 1965).

[3] See, e.g., Herbert Goldstein, *Classical Mechanics* (Boston: Addison-Wesley, 1965).

[4] Standard texts on introductory physics cover this well. See F.W. Sears, M.W. Zemansky, H.D. Young and R. A. Freedman, *University Physics*, 10th Ed. (Boston: Addison Wesley, 2000).

In 1915, with an insight that exceeded Newton, LaPlace, Maxwell, and all the others, the mathematician Emmy Noether showed that every conservation law has an underlying *symmetry principle*. Her achievement was only gradually appreciated; however, that brilliant insight has become a cornerstone of modern science.[5] Symmetry principles fundamentally **rule** all laws of physics today.

To appreciate the relationship, imagine that you're doing an experiment. First, the result should be the same whether your clock is set to standard time or daylight time. That is what "time-translational-symmetry" means, and because the equations of motion must display that symmetry, the law of Conservation of Energy necessarily follows. An extremely simple example of conservation of energy is a child on a backyard swing. Conservation of Energy pops up in countless everyday examples. Even in quantum mechanics, which has many surprising aspects, conservation of energy always holds true.

Second, a simple lab experiment conducted in Phoenix or Toronto should produce the same result. The governing equations must have "spatial translational symmetry." And that symmetry yields the law of Conservation of Momentum. Again, it turns up everywhere, including the quantum

---

[5] See, e.g., Leon Lederman & Christopher Hill, *Symmetry and the Beautiful Universe* (Amherst, NY: Prometheus Books, 2004), p. 73.

world. A home run at a baseball game can be analyzed using the Conservation of Momentum.

Third, if you examine an object or system from up or down, left or right, along any orientation, the equations must have symmetry of orientation; and from that the law of Conservation of Angular Momentum results. All figure skaters make excellent use of this principle in their spectacular jumps and spins, even if they don't know the math. A *vector* is a number that has both magnitude and direction;[6] angular momentum is a *vector* that points along the axis of anything rotating.[7] It is absolutely real, even though it's invisible. There are many counter-intuitive happenings attributable to angular momentum: a gyroscope stays balanced, a moving bicycle does not fall over, and so forth. The mathematics of handling vectors works out right every time. In the world of quantum mechanics, the special properties of angular momentum vectors determine the chemistry of the elements in the periodic table.[8]

---

[6] See, e.g., George B. Thomas, *Calculus and Analytic Geometry* (Boston: Addison-Wesley, 1953).

[7] See, e.g., S.W. McCuskey, *Introduction to Advanced Dynamics* (Boston: Addison-Wesley, 1959), p. 10.

[8] See, e.g., Raymond Chang, *Chemistry,* 3rd Edition (New York: Random House, 1988).

The existence of the law of conservation of angular momentum teaches us that being invisible and being counter-intuitive is no bar to being *real*.

Symmetry: A Cornerstone Belief in Physics

With symmetry principles thus elevated to prominence, physicists were quick to recognize that *symmetry* in mathematics is a form of beauty, which is very good. That in turn resulted in an extremely important *belief* about science: when a description of nature has mathematical symmetry, that description is correct.

The point that deserves great emphasis here is this: symmetry principles were created by God. Of course, they were *always* part of physics, even if scientists of earlier centuries didn't discover that fact. But everything (whether visible or invisible) that we term "always" was created by God. That fact frequently escapes attention. Scientists *believe in* symmetry principles without ever wondering where they came from.

Over many decades of research, several more symmetry principles were recognized and put to good use for guiding theoretical choices. The 1920s was a period when new insights occurred rapidly.[9] Because the nucleus of every

---

[9] See, e.g., Max Born, *Atomic Physics*, 7th ed. (Glasgow: Blackie & Son, 1962).

atom   is   a
combination   of
neutrons   with
positively-
charged   protons,
there   must   be
some other very
strong force that

> **Symmetry principles were created by God. Scientists believe in symmetry principles without ever wondering where they came from.**

prevents them from flying apart.[10] As quantum mechanics advanced, new particles with unusual properties were discovered. Trust in symmetry principles guided the way.

In 1930, what certainly appeared to be a violation of Conservation of Energy was countered by Wolfgang Pauli postulating an unseen particle,[11] which Enrico Fermi named the *neutrino*[12] ("little neutral one"). That theoretical prediction made in 1931 was not confirmed by observation until 1956! That shows how firmly physicists cling to belief in symmetry.

Later, in particle physics, reported violations of other symmetry principles were resolved by stepping up to a high-

---

[10] See, e.g., Harald Enge, *Introduction to Nuclear Physics* (Boston: Addison-Wesley, 1966).

[11] See, e.g., J.D. Wilson & A.J. Buffa, College Physics, 4th ed. (Upper Saddle River, N.J.: Prentice Hall, 2000), p. 943.

[12] Enrico Fermi, Translated by Fred L. Wilson, "Fermi's theory of beta decay." *American Journal of Physics*. 36 (12): 1150 (1968)

er level of symmetry that remained valid.[13] Elusive high-energy particles (quarks and gluons) combine in ways restricted by symmetry to form protons and neutrons.[14] Various quark combinations (mesons and baryons) can transform into each other as long as the governing symmetry principles are not violated. Symmetry principles have proved to be the best guide for choosing among alternate interpretations of data.

An essential point to be remembered is this: *Believing* in symmetry and "beauty" in equations is an *article of faith* among physicists.

The way we interpret all kinds of complicated data from experiments on atoms, nuclei, quarks, etc., is entirely dependent upon our *belief* in the validity of certain symmetry principles. But it's a totally reasonable belief! Looking at the equations, we find a mathematical beauty in their symmetry and say, "It just couldn't be any other way." There are no good challengers to believing in symmetry.

We can assert the validity of our beliefs convincingly, but ... only to other physicists and mathematicians. The

---

[13] See Francis E. Low, *Symmetries and Elementary Particles* (Philadelphia: Gordon & Breach, 1967), and Murray Gell-Mann and Yuval Ne'eman, *The Eightfold Way* (New York: W. A. Benjamin, 1964)

[14] F. Wilczek, *The Lightness of Being* (New York: Basic Books, 2008), chapter 12.

arguments *really are* very good, with excellent reasoning and clearly beautiful symmetry, but we must leave normal language behind and deal with vectors, matrices, and the language of mathematics.

To the great majority of people who don't understand the math, we wind up saying "trust us" – a phrase heard from high priests and gurus for thousands of years. Nevertheless, grounded in these beliefs (always linked to observation via reasoning), we have accomplished a lot. TV, lasers, medical devices like MRI – all are products of the package of mathematics that underlies physics. Physicists are so united in our belief in symmetry that our phrase "trust us" sounds very convincing.

Space – Time Symmetry

We all know that *time is real* and *space is real*. Still, there is a fundamental association of time and space, known as *Lorentz Invariance,* in physics. Space and time have to be treated the same within the laws of physics. Lorentz invariance matters most in the theory of relativity.[15] People are vaguely aware that relativity deals with how galaxies and stars formed, but they don't appreciate how exceptionally well the story hangs together. Less well known is that a cor-

---

[15] L.D. Landau & E.M. Lifshitz, *The Classical Theory of Fields* (Boston: Addison-Wesley, 1962).

nerstone of relativity is a deep symmetry between space and time.

Often people think that relativity only involves things moving very fast, near the speed of light. Much more important is the mathematical *relationship* between space and time stated by the theory. Relativity says that space and time comprise a four-dimensional *Lorentzian manifold*. The interrelatedness of space and time is characterized via 4-dimensional Minkowski space-time diagrammatic transformations.[16] Translation: this means that space and time are on an equal footing; all four dimensions (three are spatial, one is time) appear in exactly the same way in the equations of physics.

This is the *principle of covariance*. Einstein relied on that symmetry in formulating his General Theory of Relativity (GR),[17] which showed that what we think of as gravity is actually the curvature of space-time. There is excellent observational data that supports the theory,[18] but the foremost reason it is universally accepted among physicists is because

---

[16] See, e.g., Max Born, *Einstein's Theory of Relativity* (Mineola, NY: Dover, 1962), chapter 6.

[17] R. Adler, M. Bazin & M. Schiffer, *Introduction to General Relativity* (New York: McGraw-Hill, 1965).

[18] Many books cover this well. For a brief synopsis, see section 1-11 of Robert B. Leighton, *Principles of Modern Physics* (New York: McGraw Hill, 1959).

of its exquisite symmetry. The beauty in the mathematical equations is awesome.

Limited Perception

The average person's response to the basic principles of physics is simply to disconnect, attending to other topics that have a higher level of familiarity, and regarding physics as "out there." The ability to communicate is impaired by the language gap – mathematics vs. conventional words (English, Japanese, Spanish, etc.) – and without communication, people are just talking past one another.

Even when talking to each other, physicists still use human language, rooted in human experience. It is not easy for *anybody* to think of time as "just like space." The mathematical equations certainly say so, but our senses tell a different story, and that obstructs us from internalizing the idea. Even after mastering relativity, the mind still boggles at the notion of grouping space and time together. Because we have no direct human experience for reference, we cannot adjust our thinking, let alone our communication skills, to treat time and space symmetrically.

Nonetheless, the trap to avoid is thinking that God faces the same limitations that beset people.

There is only one context in which such equivalence or symmetrical transposition could *truly* occur, and that is in the mind of God.[19] God is holding the whole of history – earlier and later – in His mind. Therefore, from God's perspective there IS an "everywhen." But it's not accessible by humans; mankind is incapable of matching that understanding.

The Leap of Faith

Every time science discovers some secret of nature, we are peeling back the veil covering God's creation, inching closer to appreciating God's purpose. Some scientists will say, "No big deal, it *has to be* that way, because of the mathematics." Einstein himself was notorious for clinging to determinism and spent decades searching for "hidden variables" in physics, thinking that creation had to occur in only one certain way. He had no luck with that pursuit. Attempts to confine God within the limitations of the human mind never work out.

The human mind struggles to imagine how God sees time and space together: our mathematics works okay, but we can't put it in words because everyone's ordinary language originates from conventional human experience. To treat *time* as symmetrical with *space* requires a *leap of faith* into the realm of mathematics and symmetry principles. For

---

[19] Robert J. Spitzer, Priv. comm.

people who don't un-
derstand the mathe-
matics, the only way
to make that leap is to
"trust us."

> **Attempts to confine God within the limitations of the human mind never work out.**

A lot of people
don't want to hear "trust us" from scientists and run the
other way. They have lived okay all their lives thinking that
time is an absolute and see no compelling reason to change.
The theory of relativity is okay if it's "out there" in the galaxy,
but it becomes a very hard sell when it tries to convince
people that they suffer from a severe limitation – indeed, a
limitation that gets in the way of their ability to understand
God.

Compounding the problem, a substantial cadre of scientists have exploited their sophisticated knowledge to express
contempt for the concept of God, and hence the majority of
citizens get the erroneous impression that science and religion are enemies.

It is within the realm of metaphysical imagination that
perhaps God created just *one* very fundamental symmetry
principle, and all the others we know of today are just des-
cendants or subordinate variations of that. But such a theory
is nowhere in sight today.

Conclusion

The right way to appreciate God's creation of science is with humility,[20] conceding that we *don't know* much more than we *do know*. God can comprehend everything quite differently from human beings. The linkage between space and time contained in the theory of relativity is an example of seeing through the veil. Because we have the language of mathematics, we can transcend the limitations of conventional languages and discover a new relationship that isn't evident to the senses. To accept relativity is to perceive time as a dimension comparable (interchangeable, symmetrical) with space in the domain of mathematics.

We feel fully justified in saying that physics has *discovered* that space and time are united by this means, even though everyday experience disputes that notion. The beauty and symmetry of the theory of relativity are compelling. So, we *believe* it is an accurate description of nature.

A challenging question, seldom asked, is, "If you believe in symmetry principles, why don't you believe they were created by God?"

A precautionary principle in science says that any theory is always subject to revision, so we won't claim "certainty" for relativity; but it is definitely a *very good* theory. Believing

---

[20] Sir John M. Templeton, *The Humble Approach,* second edition (London, England: Continuum Publ. Co., 1995).

in it is quite comfortable for physicists. A theory like this, backed up by sound reasoning and observational data, is awarded a much higher status than alternate notions.

What has all this got to do with God? Of course, God understands our best human theories. So... God readily comprehends space and time in a unified way. Consequently, God's *omnipresence* applies to time as well as to space. The fact that *human* understanding has fallen off the train here is *our* problem, not a limitation upon God.

A lot is taken for granted in the way physics is customarily presented. High school physics students are taught the conservation laws, but the connection to symmetry is omitted because the details are tucked away in advanced quantum mechanics textbooks. In reality, all scientific practice is built upon a ladder of accepted theory which has its roots in consecutive layers of faith in the validity of prior science. In the next chapter, we'll look closely at how that mechanism has worked so well to bring us to the contemporary state of modern science.

Chapter 3

Can We Trust Our Senses?

The Foundational Role of Faith Within Science

Everyone will nod in agreement with the premise that there is much more to our lives than what our senses can tell us. But seldom does anyone focus on exactly what this means or ask in what way the senses are inadequate. Moreover, we are too ready to tiptoe away when someone from the camp known as *scientific materialism* takes a strident position against our perspective.

We need to ask questions about how sensory perception is limited, how measurements are made, and how real knowledge is acquired. In a Latin hymn called *Tantum Ergo*, there is a line that goes: "*Praestet fides supplementum, sensuum defectui.*" Loosely translated, this says "faith provides the supplement when the defective senses fail." Herein lies an important aspect not only of Christian belief, but of science. It turns out faith is employed all the time.

An Experiment with Eyesight

A large number of people have eyesight defects and wear glasses or lenses to correct vision. Some have better vision in one eye than in the other. Eyesight, corrected or not, does not necessarily give you full and complete knowledge of whatever you're viewing.

Take off your glasses or close one eye. With your vision thus impaired for a moment, pay careful attention to a scene: it is clearly inferior "knowledge" of what is really there compared to what you can see under better circumstances. As you do so, you realize that you are *not* seeing clearly because you remember the other knowledge gained previously at a different time and place. The experience also shows that your senses are defective and in need of a supplement.

The purpose of this little experiment is to construct an analogy for the role of faith in everyday life. The role of "other knowledge" obtained by means unrelated to direct visual observation is very important and too easily overlooked. The reality is that for the great bulk of things we say that we "know," there is an elaborate scaffolding of other knowledge – usually including faith in the truth of what some other people have said – that provides the structure with which to interpret the meaning of one particular observation.

Really Knowing

In his 1958 book *Personal Knowledge,* Michael Polanyi carefully leads the reader through the progression of steps necessary to reach the judgment that "I know" something is true.[1] One of the essential components is to place trust (faith) in the testimony of other human beings because you cannot personally duplicate every observation that has ever been done. Another component is to make the personal commitment that a proposition is *true*. There is a mixture of *objective* and *subjective* actions that must combine to produce the state called "personal knowledge." Moreover, Polanyi shows that "personal knowledge" is the kind that matters – not some abstract reality that is "out there" waiting to be discovered.

Scientific Method

Every new discovery in the entire body of human knowledge builds upon a previous collection of statements that are widely considered true by many people. The pathway by which an assertion becomes "widely considered true" requires a considerable degree of discipline, following rules which themselves came into common acceptance at an earlier time. The regression goes all the way back to some "primitive concepts" that are considered to have no need of further clarification or definition.

Many students first take note of this characteristic of the way knowledge is structured when they take geometry.

---

[1] Michael Polanyi, *Personal Knowledge* (New York: Harper & Row: 1964).

Various theorems are based upon axioms of geometry, but such "primitive concepts" as *line* and *point* are never further defined. The rules of logic, often studied in college, likewise follow paths via axioms and theorems, but terminate again in "primitive concepts."

In the *Scientific Method* of obtaining knowledge, the primitive concepts are very well hidden beneath a floor of assumptions and axioms that are virtually never questioned. In fact, for practical scientists trying to get anything done, several more layers go unquestioned: In a physics lab, an experimenter might check to see if the wall socket delivers 120 volts but would trust the integrity of the voltmeter if the measurement indeed showed 119 volts; and in any case the experimenter would not question or challenge the basic existence of electrons. In a chemistry lab, when the water faucet is turned on, it might be well to check the purity of the water, but no one doubts that the substance coming out of the faucet is mainly H_2O. Nor does anyone raise questions in chemistry lab about the existence of protons and neutrons in the oxygen nucleus.

The Scientific Method is rooted in a large collection of previous statements, now commonly agreed upon. Symmetry principles fall in that category. All this involves placing faith in the statements of prior generations of scientists. The centuries-long process that led to common agreement has been forgotten. If Galileo showed up in a high school physics lab today, he would inquire about so many "trivial" issues (such as the wall voltage) that he would be labeled "disruptive" and thrown out. To engage in science, you have to "get with the program."

Because all this dependence on faith is stored beneath the floor of the way science is done, people lose sight of it and quickly assume that today's science is totally objective. Go visit a major particle accelerator (e.g., Fermilab near Chicago) and listen to scientists talking about colliding quarks, and you'll find yourself believing in the existence of mesons before you leave that day ("meson" means a middle-weight subatomic particle). Hidden from your cognitive attention will be the dozens of acts of faith that lead upward to that state of agreement; you take for granted the "commonly agreed upon" principles of the local research team. This is simply the way science gets done, and those who challenge the status quo of prior knowledge usually find themselves standing outside in the cold.

Every now and then, someone *does* challenge the established scientific beliefs, and an upheaval occurs. The word *revolution* comes from Copernicus's book about the way the planets travel around the sun,[2] and that certainly challenged the established interpretation of science of his time. Einstein did likewise about a century ago, and in more recent years the development of quantum mechanics, electronics, plate

---

[2] Nicolaus Copernicus, *On the Revolutions of Heavenly Spheres* (Amherst, NY: Prometheus Books, 1995).

tectonics, information theory, molecular biology, DNA and so on have all resulted from occasions where someone found a better way to explain some observations than the pre-

> **A cornerstone of scientific thinking is that theory is always subordinate to *experiment* or *measurement*.**

vailing scientific ideas of the time. The phrase *new paradigm*[3] captures the notion of giving an entirely different explanation, based on different assumptions, on different choices of which "accepted" statements from the past to believe or disbelieve.

Measurements Dominate

A cornerstone of scientific thinking is that *theory* is always subordinate to *experiment* or *measurement*. As the famous physicist Richard P. Feynman once explained, no theory, however grand or "mathematically beautiful," can stand if a measurement shows that it's wrong. Moreover, one of the rules for theories is that they must be *verifiable*, meaning that for a new theory to even be considered, there must exist a means of testing it through experiment or obser-

---

[3] Thomas Kuhn, *Structure of Scientific Revolutions* (Chicago: Univ. Chicago Press: 1962, 1996).

vation. A theory that explains something no matter what the measurements show is considered no theory at all. Moreover, a theory that contains a lot of extraneous statements that (even in principle) cannot be observed is not accepted; it is either ignored or trimmed back to finite size, where it has some connection with measurable reality. The principle known as *Ockham's Razor* demands that scientific theories not be festooned with additional, unobservable claims.[4] In the scientific method, faith is given out parsimoniously.

When a scientist seeks to verify or prove a theory wrong, (s)he does so by conducting an experiment or making an observation, using a measurement instrument of some kind. It is very important that it be possible for others to repeat the measurements to check up on the claims of the first scientist. Having an experimental result repeated by another independent observer is crucial to convincing members of the scientific community to extend their faith to the initial claims. The ideal experiment is one that can be repeated by anyone at all, thus minimizing the amount of faith required by an inquiring mind. The laboratory portion of elementary science courses is intended mostly to elicit the response by the student: "Now I have seen it for myself." This pays tribute to the notion that scientific faith should only be extended cautiously.

---

[4] Franciscan friar William of Ockham (1287-1347).

Measurement methods acceptable to science have a central characteristic: sooner or later, they produce a signal that interacts with the human senses. When you can look at a clock or a meter stick or see a solution change color, that's pretty direct. Looking through a microscope at an insect demands relatively little faith – belief that the optical components (lenses, etc.) faithfully present the reality on the microscope stage. But for a great variety of modern measurements, the final connection to human eyesight is via a digitally displayed number – which appeared after several intermediate electronic steps that followed a change that occurred in a sensor connected to the experiment itself.

There are some measurements that are extremely difficult to make. As mentioned in Chapter 2, the existence of the tiny uncharged particle the *neutrino* was predicted by theory in 1931, but nobody could observe it in those days. Belief in neutrinos was maintained by an appeal to the beauty of theory, and that is always a precarious path to follow. Neutrinos were said to be all around us, passing through the earth and our bodies without interacting. A lot of people couldn't buy that notion. The cartoonist Al Capp, in the comic strip *Li'l Abner*, created a ghostly white character that was a spoof of the notion of a neutrino.

It was not until 1956 that experimental evidence for the existence of the neutrino was found, via a complex experiment that took place in a deep underground salt mine, which trusted in considerable theoretical explanation to relate the observed data to the presumed neutrino.[5] The theoretical explanation was plausible and did not affront existing quantum theory, so the connection to "commonly agreed" science was strong. Therefore, the interpretation given the measurements by the wider community of physicists was that indeed neutrinos were being observed.

Because of this and subsequent measurements, today neutrinos are fully accepted in physics. The fact that a measurement is difficult does not disqualify its underlying theory, but it is quite common for scientists to withhold their extension of faith in a theory until convincing measurements are performed.

Electromagnetism

Human senses are very defective – and limited. It is important to note that if a measurement is to connect with the senses of a human observer, sooner or later it must produce an *electromagnetic* interaction in order to have an output of the measurement device. Eyes, ears, neurons running to the brain, etc., all rely on electromagnetism for information to flow.

---

[5] C. L. Cowan Jr.; F. Reines; F. B. Harrison; H. W. Kruse; A. D. McGuire, "Detection of the Free Neutrino: a Confirmation." *Science*. 124 (3212): 103–4 (July 1956).

Communications from instruments to people, or between people, in general take place via electromagnetic interactions. There *are* other interactions in nature (the four forces are gravity, electromagnetism, the "strong" and "weak" forces), but for a person to find out about any of them requires an electromagnetic interaction. Consider the tale of Newton sitting under the apple tree and being hit on the head: the slight temporary deformation of his skull sent an electromagnetic message of pain along neurons to his brain, telling him that something had happened. Thus, *gravity* was detected via the route of *electromagnetism.*

The requirement of an electromagnetic link severely restricts the domain of what can be measured. When we look out at the stars and galaxies, we see only those that are lit up; electromagnetic radiation (radio waves, light waves, X-rays) is necessary for information about a distant location to reach us. There is general agreement that most of the universe is composed of "dark matter": things that interact with each other via gravity but give off no electromagnetic radiation capable of reaching us. Moreover, the speculation of dark energy is postulated to explain how the universe is expanding.

If we then ask, "How do you *know* that it's there if you can't see it?" the answer involves reliance upon theory – indeed, a treasured theory about how large bodies in the universe ought to behave and interact. Unless there really *is* dark matter out there, the principle of Conservation of

Angular Momentum would be violated, and nobody wants to abandon that theory, so it is considered entirely reasonable to accept the existence of dark matter. Furthermore, believing in *dark energy* is the most plausible way to preserve an aspect of general relativity.

Faith Within Science

There is a substantial component of *faith* here: adhering to a theory in the absence of observational data is rooted in the belief that the same laws of physics hold in other portions of the universe as they do where we live. It is impossible to prove this assertion, but it is a very attractive article of faith; after all, a very basic principle of science is that the universe is a rational place which is capable of being studied. The appeal of mathematical beauty and symmetry in equations is likewise very strong.

Scientists have become comfortable with beliefs of this type and don't attend to the faith component of them. "It just makes sense," "The story hangs together," and "That's the way it has to be" are some of the phrases used to justify the collection of fundamental beliefs about the laws of nature. Similarly, scientists are uncomfortable when someone opposes their belief system. "What else could it be?" is a slogan used to challenge anyone else to propose a better explanation. The experience of Copernicus, Einstein, and others shows that such challenges are occasionally victorious, but such events are rare exceptions. There is safety in the status quo today, just as there was 100, 400 or 1000 years ago.

Scientific Materialism is Self-Contradictory

The domain of thought known as *scientific materialism* holds that nothing exists except material, and all claims to knowledge other than scientific knowledge are faulty and unsustainable. This viewpoint is rooted in a failure to discern the very large component of faith inherent in every claim about scientific knowledge. The pathway to reaching this faulty state has several steps: First, agree to the articles of faith in science; second, take them for granted; third, elevate their status to the level of "axioms" of science; fourth, assume that everything else necessarily must follow the same restricted pathway of thought, and fifth, exclude from consideration anything that doesn't match this way of thinking.

Fr. Robert J. Spitzer, S.J.,[6] observed that the position of scientific materialism is self-contradictory. It says that only scientific knowledge can be real knowledge; but *that statement itself* cannot be demonstrated by science. Within a closed system of axioms and theorems like Euclidean geometry, you can establish a self-consistent system of proofs; but science is *not* such a closed system and cannot

---

[6] R. J. Spitzer, S.J., *New Proofs for the Existence of God* (Grand Rapids, MI: Eerdmans, 2010).

prove a statement *about* science. That observation is a cousin of *Gödel's Theorem,* which deals with self-referencing systems.[7] It was used nearly a century ago to undermine the philosophical system known as *Logical Positivism.*

> *Scientific materialism holds that nothing exists except material.*

To anyone who steps back and examines the entire process by which humans obtain knowledge, the arrogance of the *scientific materialism* position is evident. However, not all that many people do so. Among many religious persons, there is a tendency to just retreat in silence rather than fight back. That's a real mistake, not at all warranted by reality. The problem with retreating is that it leaves the playing field to the shrill champions of scientific materialism, who gradually become the dominant voice. Soon the peer-review system reinforces this collection of beliefs, and it becomes hard to find any articulation of other views.

There are plentiful examples from history of instances where religious and scientific views were in conflict, where in the long run the scientific view prevailed. These examples are often cited to bolster the case for the scientific method. With the hindsight of many generations, the origin of the struggle often can be traced to the "establishment" religious

---

[7] Appendix C in Stephen M. Barr, *Modern Physics and Ancient Faith* (Notre Dame, IN: Univ. Notre Dame Press, 2003).

position being overly precise, incorporating additional beliefs that didn't belong there in the first place.[8] When
organized religion assumes that God has some of the same
limitations as humans, it makes sweeping and unwarranted
statements that lead to trouble later on. Science as a way of
thinking tries to take note of its assumptions and avoid over-
stepping its bounds.

Nevertheless, science does blunder occasionally, as the
cases of *phlogiston* (a substance supposedly released during
burning)[9] and *the ether*[10] illustrate. The boundaries are not
always easy to recognize, especially if the scientific establishment supports one particular theory – which of course is
built on a scaffolding of faith in other scientists, and faith in
their interpretation of observations. It is well to remember
that Newton devoted considerable time to alchemy,[11] and
presumably swept along with him were several lesser scien-

---

[8] Giorio de Santillana, *The Crime of Galileo* (Chicago: Univ.
Chicago Press, 1955).

[9] See, e.g., *Oxford Dictionary of Chemistry* (Oxford, England:
Oxford Univ Press, 2008).

[10] See, e.g., Julius A. Stratton, *Electromagnetic Theory* (New
York: McGraw Hill, 1941), p. 102.

[11] Charles MacKay, *Extraordinary Popular Delusions and the
Madness of Crowds* (London: Robson, Levey, and Franklyn, 1852).
Online at https://www.gutenberg.org/files/24518/24518-h/24518-
h.htm

tists whose names are long forgotten. It is very easy to construct an elaborate science based on faith in an incorrect fundamental axiom or belief. Later, after the correction, science prefers to forget that it ever went down a wrong path at all. It is interesting to speculate today about what contemporary scientific beliefs will one day be antiquated – parts of biology? Psychology? Economics?

When people look back historically at cases of conflict between religion and science, it is extremely rare to see such cases examined in terms of conflicting beliefs based on different axioms. Rather, it is more common to see the tactic used in which the scientific view is labeled "rational" and the religious view labeled "superstition." Still, the often-successful exploitation of that tactic cannot change the basic underlying fact that the conflict is between different fundamental systems of *beliefs*.

Alternate Pathways

It is certainly possible to take a more balanced view of what it means to "know" something.

The scientific method provides one very important component of obtaining knowledge. Measurement really does count for a lot, but another part of the scientific method is deciding when to believe the information being stated by other people. There is widespread agreement about what it means to read an article in a scientific journal and moderate agreement about the proper set of things that should be included when writing such a journal article. To the extent that all parties adhere to those rules, information is communicated reliably, and readers are able to evaluate what they read. The anomalies that do occur usually arise when an "established theory" is undermined by new observations. Such events provide the most exciting times as science progresses.

There are several other pathways to knowledge – literature, music, poetry, art – which have been discussed at length over the centuries. The experience of living in a family conveys knowledge of several virtues in unspoken ways. The religious road to knowledge elicits different reactions from different people: some people hesitate to trust in statements by those going down that road because it is strewn with landmines associated with one or another choice of religious doctrines. Other people find fellowship and contentment sharing the journey. Mystic experiences are even less accepted because of the difficulty of communicating about the experience.

What is important in all this is to maintain a proper balance among the possible pathways. There is an element of faith in every case; in any particular group, all members share the same faith. This is just as true in science as in other areas: the faith shared by scientists is so well-accepted that it is almost never explicitly recognized, but it is still there. As stated above, it is usually "hidden beneath a floor."

There are limitations to every pathway, as well. For example, limiting one's study to the Bible alone places a boundary around one branch of Christian study and investigation. When scientists insist on the primacy of measurement over theory, a definite rule or boundary is established. Moreover, the boundaries can shift as a field undergoes advancement. Compare today with 19th-century custom: in the literary arts, today you wouldn't go to a poetry reading and complain "but it doesn't *rhyme*;" and in physics, you don't object to a particle-physics experiment that only collects data for a micro-micro-second.

The pathway of science has the major limitation that it must eventually connect with human sensory perception, and this in turn means it has to relate to electromagnetism. That sweeps off the table of scientific investigation a whole series of questions for which we can legitimately seek answers. But the rules of scientific measurement guarantee that those answers will not be found *within* the realm of science.

The reason science has gained supremacy as a basis for knowledge is that its rules of investigation – the rules about when you "know" something is true – are easy to grasp and agree with. But it does not follow that no other pathway is valid. That is the mistake made by the *scientific materialists*.

Living with Defective Senses

The "optics" experiment that began this chapter now has a clearer significance. Trying to see something without corrective lenses, or with one eye closed, is an excellent analogy for the condition of having *only* the path of science available on one's search for knowledge. If many years were to go by and you never put your glasses back on, you might forget that a better view had ever been possible.

By its very nature, scientific measurement is a "defective sense." Because measurements must involve electromagnetism, it forces knowledge of other aspects of nature (e.g., gravity) to be indirect, imperfect, limited, and dependent upon theory, which in turn depends upon faith in other people. This limitation extends to other areas of knowledge as well, far beyond physics, chemistry, and biology. In fact, assigning primacy to sensory perception places a roadblock in the path of learning about interactions between people, meaning in life, love, and many other intangibles that don't match the criteria of scientific measurements.

The proper balance merges the several pathways to knowledge. The scientific pathway is a very important one, but not the only one. It is an error to forget about or conceal the role of *faith* on any of the pathways. The defective senses always need the supplement of faith to progress along any path toward knowledge.

Chapter 4

TIME – The Falsest God of All

The opening lines of the book of Genesis have been the subject of debate for centuries.[1] It doesn't actually say "in the beginning," despite that commonplace rendition. In the original Aramaic, the very first word is *"B'raisheet"* which best translates as "with a first cause" or "with wisdom."[2] Thus, in English it would read "**With wisdom** God created the heavens and the earth." In the New Testament, the opening line of the Gospel according to John is "In the beginning was the *logos*," which is the Greek word for logic, intellect, or "word."

Verse 3 of Genesis 1 states God's very first act of creation: "God said 'Let there be light', and there was light." It was the creation of light that initiated the process of creation. This bit of physics actually is relevant to our most basic notions about God.

---

[1] S. Jaki, *Genesis Throughout the Ages* (Merrimack, NH: Thomas More Press: 1992).

[2] Quoted by G.L. Schroeder in *The Hidden Face of God* (New York: The Free Press, 2001); also in G. L. Schroeder, *God According to God* (New York: Harper Collins, 2010).

God is superior to time, a proposition with which many people would nod in agreement; but mankind has inadvertently assigned God a state of subservience to time, and by so doing has caused a lot of harm. The path toward understanding is blockaded by the very widespread (indeed, built-in) acceptance of the notion that God must obey our notion of time.

God is the one who thought up "time" and is certainly not subject to time or limited by time. To conceptualize God in a way that describes God as subject to time is to put a false god before Him. Big mistake!

The trouble is that our own understanding of life and existence is heavily encumbered by the notion of time. Our information-gathering mechanisms (the five senses) all deal in physical phenomena, the world of time and space. As we have seen, our language has a hidden dependence on time built into it at the most fundamental level: every sentence contains a verb, and verbs are usually "action" words, connoting change with time. I can't write three consecutive sentences without tripping over some implicit reference to time.

The difficulty is insidious and not easy to detect. For example, if I write "God began his creation…" – by using the word *began* I give a significance to time that God is not required to give. I can attempt to recover by adding "there was no such thing as *began* when God created the coordinate system of space and time," but then I've slipped in the time-dependent word *when*.

This is just a part of conven- tional human lan-

> **Mankind has placed a false god (time) before God.**

guage. We are stuck with time; therefore, we are stuck with *images* of God that are conditioned by time. Never confuse mankind's limited images with the much deeper reality of God.

God is not subject to our limitations; in fact, it demeans God to suppose that He might be. A variation of this, equally demeaning to God, is to limit God by thinking that our human perceptions are the only kind of knowledge there can be.

Let There Be Light

Scientists have put a great deal of effort into the field of Cosmology, trying to find out what happened at the earliest moments of creation. Observations from astronomy have been combined with laboratory experiments in high energy physics, along with theory, to reconstruct what the early universe may have looked like. The physics[3] shows that concepts like space and time lose their meaning in the extremely early universe (the "big bang"), under conditions of such high radiation density that nothing was as it seems now.

---

[3] Brian Greene, *The Elegant Universe* (Vancouver, WA: Vintage Books, 2003).

To play by the rules of physics is to concede that there can be no such thing as time "before" that initial cataclysm.

God had a much better view of the situation. God's first creative act was to say, "Let there be light." Why was light the first thing God created?

It is first necessary to bring up a physics principle: the speed of light is what makes time different from space. Mathematically, space and time are not different at all – they're just mathematical dimensions. The four dimensions appear on an equal footing in all the equations of physics. The space dimensions $\{x, y, z\}$ are commonly measured in meters, and the time dimension t in seconds. The speed of light is denoted by C and measured in meters/second. ($C = 300,000,000$ meters per second $= 3 \times 10^8$ m/s.) Multiplying C times t gives dimensions of meters, so the "units" come out correct in the equations. This is well-established science.

One basic feature of dimensions is this: dimensions are orthogonal (perpendicular) to one another; the term is explained in Appendix A-1. There is nothing special about a multiplicity of dimensions. It's pretty boring mathematics to have zillions of unrelated dimensions; you might call it a "formless void."

To create light, God drew out of the infinite manifold of possible dimensions a small subset of dimensions (four, we think),[4] and configured them into a type of complex relationship. God made all four perpendicular to each other. To visualize this in a graph, we designate the time dimension as being in the *complex plane*, in a direction perpendicular to the *real number line*.

The customary representation of complex numbers uses the mathematical symbol *i* to label that perpendicular axis. Therefore, the extra time-dimension isn't merely C times *t*, but is *iCt*, where *i* denotes the "imaginary number." That's what makes the dimension of time distinct from the three spatial dimensions. Unfortunately, the terribly faulty word "imaginary" has become standard to describe that axis, leading people to think there is something weird about it.

When the complex plane is introduced in trigonometry, we learn that the value of *i* is simply the square root of minus one, which facilitates mathematical rotations in the complex plane. When one dimension is related to three others via complex numbers, all sorts of things can happen.

In the language of physics, we say the *symmetry* was *broken* between dimensions $\{x_0, x_1, x_2, x_3\}$, leaving $\{-iCt, x_1, x_2, x_3\}$. Deliberately breaking a symmetry is a creative act. "Let there be light" is equivalent to saying, "Let space be different from time, and let there be a relation between them." God

---

[4] It is separately probable that our existence involves still more dimensions, but these are not at issue here, and don't affect our focus on "time".

created the *relationship* between space and time and, thus, defined a real coordinate system. There could be no such thing as time in the absence of a *relationship* to space (a speed of light). It is only because there is light that "space" differs from "time."

Initial Creation

We concede that the very first instant is known to God, but certainly not to humans. What happened in the extremely early universe is beyond the scope of contemporary physics. There is a shortest length (the *Planck length,* $\sim 10^{-34}$ meters) and a shortest time (the *Planck time,* $\sim 10^{-43}$ seconds) below which our concepts and equations don't work anymore because of the *uncertainty principle* of quantum mechanics. The density of the initial burst was so great ($> 10^{98}$ kg/m^3) that space and time were indistinguishable, not yet emergent. This period is known as the pre-*Planck era* or the *Quantum Gravity* era or the *fundamental level.*

Many attempts have been made to explore that region with the aid of advanced mathematics. One promising method is the use of *noncommutative geometries* and algebras, by Michael Heller and colleagues.[5] Normal concepts like "probability" and "causality" are not what we expect. A

---

[5] Michael Heller, *Creative Tension* (Radnor, PA: Templeton Foundation Press, 2003), chapter 9.

favorable aspect of that treatment is that when cooling and expansion proceed enough to reach the Planck time and Planck length, certain limiting conditions apply; space and time are distinct, and standard physics emerges.[6]

Heller is cautious not to claim too much: "To claim that God uses a noncommutative probabilistic measure in designing the universe would display an equally enormous naivety." That's definitely a comfort to everyone who cannot penetrate those upper reaches of mathematics. We are content to know that God can easily do so.

All this is a very different interpretation from what people ordinarily imagine upon hearing the Genesis line, "Let there be light." Note that God didn't just create an electromagnetic field (light) and place it into a pre-existing realm of space and time. There was no "pre"-existing anything because time can only exist in a relationship with space.

In everyday life, time is obviously real to us.[7] But *we* have come on the scene long after the initial creation. Unlike inanimate rocks, we "experience" time being different from the other three dimensions. Mathematically, however, it began as just one entity. The "arrow" of time *emerged* as the

---

[6] Heller, *Creative Tension*, chapter 10.

[7] For a clear exposition of the seven ways in which time is unidirectional, see pp. 183-186 in J.M. Templeton & R.L. Hermann, *The God Who Would Be Known* (New York: Harper & Row, 1989).

initial intense radiation field cooled and condensed into particles in the very early universe.

Without the relationship that light implies, x_o has no special properties, and it is not particularly different from the spatial dimensions. God's act of creating light initiated both space and time, a distinction that only has meaning *after* God created light.

Time came into existence with the creation of light. There is no such thing as "before" the creation of light.

Subsequent Creation

What happened next? (Note that now it's okay to use time-like words.) In the subsequent development of the universe, additional symmetries were broken; gravity quickly separated from the other forces; within sub-micro-microseconds, strong, electromagnetic, and weak forces separated. It is fun to look for the guiding hand of God in those other symmetry-breaking events; verse 4 of Genesis 1 says that God separated the light from the darkness, still on the first day.

However, our focus here is on time. Appendix A-2 reminds us that God readily manages the swift timing of events that astound human beings.

All that rapid diversification is well-described in books on early cosmology.[8] Shortly thereafter, the universe took on the physics that has evolved only a little more to the present day.

Once Genesis gets rolling (verse 6), time plays an essential role in *what one can write about.* The subsequent text of Genesis 1 describes a sequence of creative acts by God, and, of course, a sequence implies time passing. There is no mention of God's creating a coordinate system (space-time) because neither scribe nor reader had any ability to think about there **not** being a coordinate system. "Formless void" would have to do. Since around 1985, we would probably select the newer word "chaos."[9] For each step, there is a transition from chaos to order,[10] embedded in the words, "It was the evening and the morning of the [...Nth] day."

Scholars generally agree that "world" in Genesis means "the universe," not "the planet earth." They have debated the exact or metaphorical meaning of terms like "day," "formless void," "dome," and so forth, but nobody scrutinizes the phrase, "Let there be light." I think it is appropriate to take "light" quite literally.

---

[8] See, for example, Steven Weinberg, *The First Three Minutes* (New York: Bantam Books, 1979).

[9] James Gleick, *Chaos* (New York: Viking Books, 1987).

[10] G.L. Schroeder, *The Science of God* (Glencoe, IL: Free Press, 1997).

With the creation of light, time came into existence. The word "before" has no meaning where there is no time. All words referring to time only have meaning when time is already in place. It could not be any other way. Had Genesis begun, "God used to live in this neat formless void, see; and then He...", we would have to conclude that either God was subject to time, or that the human writer of the text was not particularly inspired.

It is fair to ask, "Why didn't God say so when He inspired the Bible?" Here it is well to remember that man can only handle so much, and human language is so intertwined with time that it was beyond the ability of man to grasp such ideas in those days. That was the condition on the receiving end of Genesis. And if the prophets couldn't quite catch God's meaning correctly, how much worse off were the translators? At the end of the 19th century, time was still assumed to be an absolute. It awaited the insights of Einstein in the 20th century for the relation between space, light, and time to enter human consciousness.

Without even the slightest hint of a theory of relativity, about 1600 years ago St. Augustine got it right.[11] He said that time was created by God, and then refutes some common misconceptions:

---

[11] St. Augustine, *The City of God,* Book XI, section 6 (New York: Random House, 1950), p. 350.

"Since   then,
God, in whose
eternity   is   no

About 1600 years ago, St. Augustine got it right.

change at all, is the Creator and Ordainer of time, I
do not see how He can be said to have created the
world after spaces of time had elapsed, unless it be
said that prior to the world there was some creature
by whose movement time could pass...

...the world was made, not in time, but simultaneously with time. For that which is made in time
is made both after and before some time – after that
which is past, before that which is future. ..."

In Book XI section 7, St. Augustine goes on to state that
God created light before the sun and says that kind of light is
"beyond the reach of our senses." *The City of God* also
recognizes a similarity between space and time in that St.
Augustine says that it makes no more sense to ask why God
was idle for a while than it does to ask about other possible
locations for the creation.[12]

Pretty good for a guy in the 4th century!

The point asserted here should be distinguished from
what St. Augustine said. In the 1600 years that have followed
St. Augustine, science developed in many ways, not least the
theory of relativity. The notion of a manifold of dimensions

---

[12] St. Augustine, *City of God*, book XI, section 5.

(mathematically, perhaps an infinite number of dimensions) from which four are singled out with a relationship among them $\{x, y, z, iCt\}$ is totally beyond the thinking of that age. St. Augustine wrote as though eternity is merely unchanging *conditions*, even if time marches on.

My assertion is that God's specific creative act was to break the symmetry among these dimensions, and that "Let there be light" defined the special relationship that turned *space* and *time* into separate concepts. In this way, I hope to emphasize that *time* is distinctly subordinate to God. This act of symmetry-breaking conforms fully to St. Augustine's concept of creation. Certainly, I agree with St. Augustine's point that certain questions are meaningless; but I also draw attention to the limitations that human language and experience place upon our ability even to discuss God's creative acts.

How People Grasped "Time"

Over many centuries, the immutability of time in people's lives evolved into a certain respect, indeed awe, toward time; and hence when mechanical clocks came along, it wasn't long before the image of God as "the watchmaker" became prominent. This led to the suggestion that God once wound up the watch, started it running, and never looked at it again. This is known as the *Deist* viewpoint. This is a very unsatisfactory image of God, one that led many scientists to a *de facto* atheism, or at least a "God does not matter" outlook. Most certainly it left God distinctly subordinate to time, only able to tinker with His universe in certain limited ways. That was not too bad as long as time was held to be absolute, and so religious leaders put up with the image and eventually accepted it. But what happened when time lost its pre-eminence?

The dominance of clocks and time evaporated in the 20th century. In *The Ascent of Man*,[13] Jacob Bronowski explained light and time via this image: While riding on a trolley car, look back at a clock on a tower and ask, "What if I could ride on the beam of light leaving the hands of the clock? Would any time pass?" The answer is no, and Einstein saw that time does not pass when you're traveling on a light beam, moving with the speed of light.

Today, most physicists have accepted that time is not absolute, but among the general populace, nearly everyone still holds an image of God in which God is subject to time. There are dozens of ways that humans have allowed time to distort their understanding of God. Limited, incorrect theological images are everywhere. Here is a single illustrative example of this phenomenon:

Heaven: For over a century, we have abandoned the idea that heaven is a place; we smile at the image of sitting on a

---

[13] Jacob Bronowski, *The Ascent of Man* (New York: Little, Brown, 1973).

cloud strumming a harp. We have faith in God that He has something much better in mind for heaven. However, by repeating the word *afterlife* so often, we have given an undeserved primacy to time. An "after"-life heaven is visualized as merely an extension of our physical world, quite inferior to the full reality that God knows. God has something better in mind. We need to take an even bigger leap of faith and concede that heaven is not a *time*, either. This will be discussed more in Chapter 12.

Consequences

The equation of general relativity treats space and time symmetrically. Little by little, word of this is getting around, and mankind's comprehension is advancing. Unfortunately, there are many religious leaders who recoil from this – perceiving it as an "attack." It should be exactly the other way around. Relativity invites the religious person to reassert God's dominion over time.

Any image of God that is confined by time is too limiting; similarly, any image of our ultimate *relationship* with God that is dependent upon time is doomed to fail. We have to start imagining a totally different kind of existence. (Stanley Jaki speaks of "spiritual *dimensions*."[14]) This is very difficult

---

[14] Stanley Jaki, *The Relevance of Physics,* (Chicago: Univ. Chicago Press, 1970).

to do because our entire language, culture, and thought-structure are all conditioned by our customary understanding of time. Thus, our images drawn from everyday life are not going to be much help.

Conclusion

The barrier identified in this chapter is the natural human tendency to assume that time is immutable and everything is subject to it, including God. Acknowledging a limitation is the first step forward. We can realize that there are qualities of God for which a human treatment of time is insufficient. With that misconception about time removed, we can advance beyond past dilemmas.

If there is one cornerstone principle in anyone's religious faith, it is that *reality* is more than meets the eye. The *Materialist Superstition* says, "If you can't measure it, it doesn't exist." The religious person will not agree that this physical world is all there is. Notice, however, that the standard human image of God's being restricted by the time constraints of the physical world contradicts cornerstone beliefs. When we say that God is almighty, we are saying that He has power over *all* reality, of which the physical world (including time!) is only a subset. Certainly, He can create a mode of existence that grandly exceeds the physical world.

It's difficult to criticize a set of beliefs that most people naturally accept, including the "absolute" supremacy of time. Many people lead good lives and reach heaven without ever hearing of space-time, let alone understanding it. After all, time is with us constantly; we can hardly speak about anything without implicitly invoking time.

Nevertheless, these limited perceptions have endured too long, and change is overdue. In the present age of modern science, including relativity, it is anomalous that people still think that God is constrained by time. For example, God's ability to answer prayers is not restricted by the time-sequence that humans are accustomed to – but few appreciate that. This lapse in understanding of God's power arises from assuming that human limits apply to God. The way to rise above that image is to acknowledge that God is the *transcendent* Creator of time, independent of and external to it, and exempt from its constraints. With that expanded image of God, humans could more readily *trust* God to act in their best interests.

Chapter 5

Something More Beyond Space-Time

The human desire to know *more* is never fully satisfied. The experience of "now I understand better ... but still not quite perfectly" is familiar territory. Everyone wants to advance to higher stages of understanding. We spend our lives striving for something "other" or "more," without knowing exactly what that is. Wondering and questioning underlie the human insistence that there *is* something more, something beyond the world we perceive.[1] But a satisfactory understanding is elusive.[2]

There are many ways in which our constraints are obvious: Every human has a natural fear of death; our lives are ruled by *time,* and so forth... That leads one to conclude that there must be some barrier in the way. From the expressions of mystics to the formal logic of philosophers, over the millennia a wide range of partial explanations have been provided. *Every* explanation is necessarily presented in *some*

---

[1]   Robert J. Spitzer, S.J. *The Soul's Upward Yearning* (San Francisco: Ignatius Press, 2015).

[2]   Sir John Polkinghorne, *The Faith of a Physicist, Gifford Lectures* (Minneapolis, MN: Fortress Press, 1996).

language and, hence, is limited by the boundaries of a language.

Visible and Invisible

One very familiar limitation is this: the words on the page in the Bible often don't convey the full meaning. The statement by Jesus "My kingdom is not of this world" leaves an obvious question unanswered. When the disciples asked Jesus why He always taught using parables, they at least grasped that He must have been talking about "other" or "more." The Christian faith has retained that principle from the outset. It is embodied in the Nicene Creed, which begins: "I believe in one God, the Father Almighty, Maker of heaven and earth, and of all things visible and invisible."

Many centuries later, today the words "all things visible" generally refer to everything that is accessible via scientific instruments. The microscope, the telescope, X-rays and ultrasound have greatly extended the range of our senses. When something is made from quarks, particles, atoms, or molecules, that's part of "all things visible." Things that exist in space and time (even distant galaxies, and even "dark matter") are included because knowledge about them, derived from observations, falls within this category. The laws of physics that describe what happens within space and time are likewise associated with "all things visible."

But what about that "invisible" creation? The Nicene Creed contains an important commitment on the part of the Christian believer: that God created *more* than just the world we see; that there *is* a lot more to which we must attend. *And it's of a type that will elude our scientific instruments.*

The Space-Time Continuum

As a preliminary to distinguishing between "visible" and "invisible" creation, it is helpful to look closely at one aspect of the way we perceive things via science. Comparing past and present understanding provides insight into the way human thinking advances.

In contemporary physics, we speak of the "space-time continuum," referring to the four-dimensional manifold made up of three spatial dimensions (x, y, z) together with time (t). The laws of physics contain a mathematical symmetry among these four dimensions, and the spatial dimensions are interchangeable with the time dimension via relativity, as we saw earlier. In fact, it is a requirement upon any new proposed theory that it must contain such symmetry and equivalence of dimensions, or it will be dismissed.

To grasp the idea of a "fourth dimension," think about the differences between a two-dimensional image of reality (such as a movie) and the three dimensions (including depth) that everyone experiences. If your depth perception is faulty, you don't "see" the third dimension the way most

people
do, but
you learn
anyway
from
other
people

> **The Nicene Creed contains an important commitment on the part of the Christian believer: that God created more than just the world we see.**

that it is real. In an analogous way, our customary intuition of spatial relationships does not give us the "depth perception" required to understand that time is the fourth dimension. We must *learn* that fact.

That way of comprehending physics is barely one century old. In the days of Isaac Newton, around 1700, time was certainly not considered a "dimension," but was thought to be absolute and immutable. His equations of physics described how objects in space moved *within* time. Newton's achievement greatly advanced science, and classical mechanics remained dominant even into the 20th century. Today, we look back upon Newtonian physics as an approximation to our more comprehensive laws of quantum mechanics, general relativity, etc.

However, a whole lot of important developments in thought took place during the centuries of Newtonian dominance. Foremost, there was no reason to think of "time" as having any similarity to or association with "space." Newton's picture corresponded perfectly well with the ordinary experience of human beings, in which space and time are perceived as entirely different. From a child's earliest perceptions onward, a strong distinction between space and time is embedded in our thought structure; it's reinforced by every culture and language.

General Bias and Scientific Progress

In mid-20th century, in his book *Insight*, the philosopher Bernard Lonergan examined the processes of the mind by which knowledge is reached.[3] Among his many other contributions, he pointed out that there is "bias" that impedes a person's ability to grasp realities beyond one's own limits. Lonergan denoted "individual bias," "group bias," and "general bias." For example, group bias is when everyone you know believes the same things, and it only gets corrected when someone from outside enters and disrupts the prior beliefs. Lonergan explains the condition of *general bias,* in which the entire populace (say, a country) suffers from the same erroneous perception,[4] and there is no pathway or motivation to correct it. When there is *general bias* occurring, the entire society adheres to one specific way of thinking. Lonergan discusses examples of such group bias, such as Marxism and other totalitarian systems.

---

[3] Bernard J. F. Lonergan, *Insight, an Inquiry into Human Understanding* (New York: Longmans-Green, 1957).

[4] Lonergan, *Insight,* chapter 7, section 8.

General bias does not just pertain to social and political frameworks of thought. Scientific progress can likewise be impaired. Go back 5000 years and look around you: the world is obviously flat. Later, Pythagoras and other early Greek mathematicians did measurements that showed it's a sphere, but that knowledge spread only very slowly. In Galileo's day, opponents of the Heliocentric system controlled all the institutions of society; many leading authorities refused to even look through Galileo's telescope. That general bias was dominant across Europe.

The centuries that followed Galileo and Newton included the Enlightenment period, and naturally all those philosophers perceived time as Newton did. However, toward the end of the 19th century, inconsistencies began to appear, and experiments showed that Newton's classical picture could not be entirely correct. Using mathematics where space and time were placed on the same footing, motion could be better understood. Thus, a *new paradigm* was introduced, and physicists began the unification of space and time. In the early 20th century, Einstein introduced relativity, wherein space and time appeared in an entirely symmetrical way. By 1920, there was convincing experimental verification of Einstein's theory.

Ever since, the four-dimensional representation of *space-time* has been standard in physics, and with suitable notation the equations look alike, which underlines the symmetry. But writing something mathematically is by no means the same as giving a verbal explanation of what it means. That is much harder to do because of the constraints inherent in language.

Scientists look back and ask, "How was the general bias [about time being totally different from space] overcome?" The answer is "mathematically." The new insight required trusting in the power of mathematics to represent reality better than what could be accessed via the senses. Because physicists *believed* that symmetry principles are at the foundation of physics, it was possible to choose among multiple possible mathematical pathways. Today, we look at relativity theory and say in hindsight "of course!" because of its very beautiful symmetry.

That wasn't the case a century ago; there were many others (not only scientists, but also philosophers, churchmen and persons in disciplines far from mathematics), who resisted the notion that space and time were in some way interchangeable. It just didn't square with everyday human experience, and hence there was resistance to thinking in such terms. In the same way that Lonergan described the usual suppression of statements that violate the general bias, for several years Einstein's work was ridiculed and denounced.[5]

---

[5] Fortunately, the top Nazi leaders were scornful of "Jewish physics" and didn't fund an all-out program to quickly build an atomic bomb. The *Manhattan Project* was not similarly limited.

Today, with a globe in every elementary school classroom, everyone realizes the earth is a sphere, and the "flat earth society" is mentioned jokingly. But Einstein's innovative synthesis of space and time has not caught on very well, even after a century. The way that people have always thought about time, based on everyday experience, is an example of general bias.

To this day, the vast majority of people have no need to think about the symmetry of space and time, or to pay any attention to relativity. Even the astronauts could get along with only tiny numerical errors by using Newton's classical mechanics.

Nevertheless, the general bias has been breached, and the 21st-century understanding of nature places space and time in a symmetrical relationship. Although we look around and still *see* only dimensions *x, y* and *z* (back-and-forth, sideways, up-and-down), through mathematics we have been able to adapt our understanding to go beyond sensory perception and think of a four-dimensional concept that includes time as one dimension.

Knowledge has advanced, concepts have expanded, and we understand somewhat better the "all things visible" part of creation. Mathematics has been essential to that advance. We got there by stepping up to a new level of thinking, where an additional dimension (beyond space alone) was recognized.

Other Mathematical Advances

Within some sub-fields of science, the use of trans-formations into other dimensions has led to very fruitful theories and many applications. For example, solid-state physicists have worked with "reciprocal space" for decades,[6] using not [spatial coordinates & time] but [momentum coordinates & frequency]. The results are excellent: everyday electronic devices such as cell phones, TVs, and computers are rooted in the validity of that theory. In "reciprocal space," we *can* imagine higher dimensions, and manipulate them mathematically without being limited by constraints of conventional language; hence, using additional dimensions has become routine. As long as it does not cause a conflict between theory and measurements, expressing concepts in alternative abstract dimensions is okay.

Higher-dimensional thinking works so well in the physical sciences that it is natural to seek ways to adapt that approach to issues relevant to humans. The constant risk in doing so is that some additional implicit assumption will slip in unrecognized and lead to a seriously distorted outcome.[7]

---

[6] Charles Kittel, *Quantum Theory of Solids* (Hoboken, NJ: Wiley, 1963).

[7] NB: There have been past attempts to apply dimensional thinking to express ideas about humanity, dating back to the ancient Greeks. The 17th century mathematician and philosopher

The Lesson in the Story of *Flatland*

Detecting *individual bias* within yourself is hard to do; and *group bias* is only exposed when someone comes from outside the group who is free from that bias. Without employing Lonergan's terminology, a delightful little novel was written in 1872 by Edwin Abbott,[8] who was the headmaster of a boys' school. It was the story of a person (Mr. A. Square) living in a two-dimensional plane (*Flatland*), and his encounter with a sphere who visited from a higher-dimensional reality. The sphere was *not* limited by the group bias of the Flatlanders.

The humor comes in seeing Mr. Square's struggle to learn about a reality beyond his experience and comprehension. The story is synopsized in Appendix B. The important message for the human reader is to realize that the three-dimensional world we experience might very well be less than our true reality.

Leibnitz lived at a time when *maximization principles* were fashionable in physics, and he pursued that implicit assumption. Leibnitz constructed an expression of God creating a world in which the "good" was maximized. That led subsequently to Voltaire's derision of "the best of all possible worlds" in the novel *Candide*. Ever since, people have been wary of using any type of mathematical description of philosophical concepts.

   [8] E. A. Abbott, *Flatland*, 6th edition (Mineola, NY: Dover, 1952).

There are certain subtleties about Flatland, not stated by the author:

When you have *access* to higher dimensions, you have *control* over lower dimensions. The reason the sphere can freely enter and leave two dimensions is because he is three-dimensional, and hence anything he does in two is only a *projection* of his own higher reality. The sphere doesn't need all his dimensions at once to interact with Flatland, which gives him an extra "degree of freedom."

Without access to higher dimensions, you forfeit control over lower dimensions. A. Square could not have made a two-dimensional replica of himself; there is no shadow *on* the plane of a creature *in* the plane. Mr. Square is not of sufficient dimensionality to have control over two-dimensional representations, so he couldn't even have a photograph or a two-dimensional signature. We are of higher dimensions and hence we *do* have control; moreover, we are smart enough to realize that we have such control over fewer dimensions.

One significant point worth emphasizing is this:[9] The drawing of Flatland on the printed page is not just a *picture* of A. Square's world; that *is* his world. We higher-dimensional readers have sufficient control over that world to duplicate it at will, by simply printing more copies of a page.

---

[9] Thomas F. Banchoff, presented at *Cosmos & Creation* (May 25, 1996).

The ambitious reader who writes an alternate version creates another Flatland just as real as Abbott's original.

> **When you have access to higher dimensions, you have control over lower dimensions.**

Mobility Comes from Control

We humans can travel on the surface of the earth. Simply by walking around, we can occupy a position $\{x, y\}$ over and over… as many times as we like, although at sequential times. We can easily execute virtually the same trajectory $\{x(t), y(t)\}$, perhaps by taking the same route to work each morning – again, the only distinction is that these occur at sequential times.

We can occupy several different positions $\{x_1, y_1\}$, $\{x_2, y_2\}$, $\{x_3, y_3\}$… at the same time by placing *projections* of ourselves at those points. At sunrise, you may be able to cast a shadow 200 yards long; then, your projection (shadow) occupies a large range of positions $\{x_j, y_j\}$. Run for political office and plaster campaign photo-posters on every telephone pole in town; your two-dimensional image will occupy many discrete positions $\{x_k, y_k\}$, all at the same time.

Moving up to three dimensions, humans can also vary our z-coordinates, say by jumping up and down, or by getting in an airplane, and later returning to the same coordinates $\{x, y, z\}$. Again, all such recurrences occur at sequential times. What we are unable to do is place many three-dimensional representations of ourselves at various coordinates all at the same time. We face the three-dimensional analog of what Mr. Square experiences in two dimensions. Also, we have gotten used to this limitation: people humorously complain about being unable to be two places at once.

Anticipating the next chapter, it is important to note that our ability to occupy the same point $\{x_e, y_e, z_e\}$ again and again (say, by coming home at night) is meaningful only because of the existence of memory. If we did not have memory, it would be impossible for us to distinguish how freedom of movement around the three spatial directions takes place as time passes.

Conclusion

The invisible side of creation is so much more than what we perceive via our senses. It is essential to recognize that thinking only in space-time terms has imposed an artificial limitation. Thought patterns, culture, and language get in the way. It took a very long time to overcome the *general bias* of believing time was absolute, but now we understand time as a dimension akin to space. We have been set free to begin thinking at a higher level, in additional dimensions.

The value of *Flatland* is that it presents an example of meeting an intelligence far greater than our own. At the same time, it shows the reader one way to imagine a higher dimensional life. It invites us to expand our thinking upward toward greater realities, beyond what is accessible via sensory perception.

Obviously, God is one extremely important reality in that category. Eager to learn, humans are immediately inclined to ask questions. One such question is this: If God is transcendent and stands outside of time, how does God see time?

I don't know; I'm about as qualified to discuss *that* as Flatlanders would be to discuss the interactions between colliding spheres. In fact, the way in which I'm unqualified is entirely like their disqualification: I don't have access to the extra dimensions needed to encompass the topic. I'm not able to stand outside of time the way God can.

But we need not give up. We acknowledge that access to the entire spiritual world requires a lot more than sensory perception. Therefore, we utilize science as a signpost that points in the right direction. Science and faith are parallel paths toward the beauty, goodness, and truth that are manifestations of God. Their teamwork has been described as a *Great Partnership* because of the way they pull together toward the same goal.[10] That immediately reinforces the importance of Scripture, a point that the ancients grasped

---

[10] Jonathan Sacks, *The Great Partnership* (New York: Random House, 2011).

rather well.[11] Many people have found that an attentive and contemplative reading of Scripture is the very best way to elevate their thinking – even if they've never heard the term "higher dimensions."

---

[11] Stephen M. Barr, *Modern Physics and Ancient Faith* (Notre Dame, IN: Univ. Notre Dame Press, 2003).

Chapter 6

Exploring Higher Dimensions

To look beyond the limits caused by language, this chapter borrows a notion from the field of mathematics: it deals with the concept of higher dimensions. It recognizes that the realm of God's creation goes far beyond our human ability to express ideas; and that customary means of communicating are an impediment.

The "language" of geometry and mathematics permits an exploration beyond the usual bounds, beyond what can be visualized. Dimensional thinking can facilitate an upward step toward understanding our relationship to God.

Deficient Human Models of God

For a believer in God, it is a very short step to acknowledge that if humans can imagine many more dimensions, then God can certainly do at least as well. The opening words of Genesis include terms like "void" or "abyss." (In some translations we find "In [the primordial substance] there is potential.") It also says, "The spirit of God was stirring above the waters." All this could just as easily be visualized as an infinite set of dimensions available to God. There is no reason to confine our thinking about God to fewer dimensions.

Every such constraint really only expresses a boundary of human thought, not a limitation of God.

As discussed in chapter 4, a very natural error for man to make was to think that God somehow exists *within* time, because all of us do. We have no experience of time "not being there," so we are unable to construct an image of *anybody else* existing independently of time. Newton's formulation of physics placed everything (including God) *within* time. Every Enlightenment philosopher (Hume, Nietzsche, Rousseau, Voltaire, etc.) over the ensuing centuries accepted the very same error and constrained their images of God to match that restriction. It is not surprising that so many of them disbelieved in such an inferior god.

The mistake was certainly not obvious, but still, it had damaging effects. Recognizing and overcoming this mistaken perception about time is definitely an advance in our human understanding of God. But it is impossible to go back 3 centuries and repeal all the theology that was constructed based on *time*'s having supremacy. It's too deeply ingrained in everyday human experience and thought processes. Rather, what should be the path *forward*?

If we go back to square one, finding God simply *present* to an infinite number of dimensions, then He could easily carve out four of them "for starters" and create space and time together, as indicated by St. Augustine. In doing so, God's affinity or accessibility to all the other dimensions is in no way compromised. (In fact, going back to square zero,

God originated logic, symmetry, mathematics, and the very concept of dimensions.) He

> **A very natural error for man to make was to think that God somehow exists within time, because all of us do.**

can utilize additional dimensions in ways that include linkages with the dimensions of space and time. That freedom is quite different from the restriction and subordination to time that was inherent in a Newtonian worldview.

Higher Dimensionality

Naturally, we would like to know "what are all those other dimensions?" Although I, too, am a captive of human language, thought, and culture, I can certainly say – aided by the language of mathematics – that there exist realities lying beyond atoms and molecules in dimensions that lie beyond space-time.

Mathematical physicists who study very tiny particles using *String Theory* have employed additional dimensions for decades,[1] but those tiny dimensions are said to be hidden or *curled up,* inaccessible to measurements. That's not what

---

[1]   Brian Greene, *The Elegant Universe* (Vancouver, WA: Vintage Books, 2003).

I mean. Instead, I'm contemplating additional dimensions that pertain to human beings.

The distinguishing characteristic of each dimension is that it is *orthogonal* to all other dimensions (see Appendix A-1). "Space" was formerly limited to only x, y, and z, and it's easy to visualize orthogonality, which means the same as *perpendicular*. Prior to relativity, hardly anyone applied the word "dimension" to time, but now we do. That advance enhanced the clarity of our thinking, of our model of the universe, and we have beautiful and symmetric equations of physics to express that clarity.

The merit of using the word "dimension" may not be apparent; it conveys "math" to lots of people and causes some discomfort. The term "degree of freedom" is preferred by some and can be used interchangeably. The point to be kept in mind is that each additional dimension/degree is totally different from all previous (lower) dimensions.

The notion of higher dimensionality is not new. Long ago, the Greek mythology about Icarus and Daedalus addressed the desire to enter a higher dimension. In *The Republic*,[2] Plato stated the analogy of the cave, wherein the life that we experience is like a shadow (a projection downward to fewer dimensions) of a higher reality. Throughout history, a great deal of religious expression and literature has utilized the notion of higher dimensions.

---

[2] Plato, *The Republic*, Book VII.

We take air travel for granted and forget that until the 20th century, we couldn't get off the Earth's surface. Under that circumstance, someone wishing to express "other-world" concepts would simply refer to the z-axis: Heaven was "up there," meaning a place you can't get to. The imagery was used so extensively in paintings, prose, and poetry that a literal interpretation became the norm for centuries. But all the while, the underlying concept was "beyond" or "more," not merely "up."

That isn't just an analogy. It expresses a condition that must be overcome with each new upward step beyond conventional experience. A long time ago (pre-hominid days?), there were no abstract concepts, no words to describe them, no numbers, etc. As civilization advanced, countless new phenomena introduced new concepts, which extend into new dimensions. We take *reasoning* for granted, but it wasn't always so; such intellectual activity was once a novel innovation.

One good example of increasing dimensions comes from the very creative thinking of Teilhard de Chardin.[3] Teilhard extended the concept of energy, adding a component of *radial* energy that was *orthogonal* to conventional kinetic

---

[3] P. Teilhard de Chardin *Le Phenomene Humain*. The book was translated into English as "The Phenomenon of Man" (New York: Harper & Row, 1961) and later as "The Human Phenomenon" (East Sussex, England: Sussex Academic Press, 1999).

and potential energy. That radial energy reached outward and upward into a new dimension. Sketches of that expanded concept of energy are shown in Appendix C.

Accessing More Dimensions

The questions naturally arise: "Are humans confined to only the four space-time dimensions in which atoms and molecules are found? Can we become involved in higher-dimensional reality somehow?" The adherents of *nihilism, materialism,* and *scientism* would promptly answer "No, we can't" and exit the conversation at that point. Their belief structure ensures that the only part of creation they perceive is the part accessible to science, the "visible."

At this point, the Christian affirmation that God created "all things visible and invisible" makes an enormous difference. To us, it is obvious that God *can* do whatever He likes with additional dimensions, so He presumably has done so. Our enterprise is to follow the upward steps that lead to higher-dimensional reality.

We can recognize some of these higher realities easily. The well-known "ladder" or "hierarchy" of disciplines goes: math → physics → chemistry → biology → behavior →...→ music, art, etc. The increasingly higher levels are truly different from each other and are distinct branches of advancing humanity. At each new stage, there is something new added, e.g., living systems are more than just chemistry.

The upward march to successively higher levels approximately corresponds with the advance from a bare rocky planet to modern civilization. The ladder keeps going up further. Spiritual qualities like love and mercy and loyalty are yet more advanced functions.

When climbing back down the ladder, something is deleted at each stage. Characteristics like eyesight, hearing, memory, language, abstract thinking, intelligence, and so forth – all familiar concepts – cannot be reduced to elementary levels. The futile attempts to reduce human culture to genetics expose that truncation process.

Here I associate these advancing faculties with additional dimensions beyond the four included in space-time. I am unable to specify any one-to-one correspondence, nor can I assign labels to dimensions. But I am fully confident that human life greatly exceeds the minimal dimensionality of space and time. In the progress of mankind, there have been many consecutive upward transitions over eons. As a result, the human being is now positioned at a level somewhere in a multi-dimensional space. It is reasonable to expect still further upward transitions in the future.

Higher Rungs on the Ladder

To illustrate such a transition, a Biblical example may help. In the Garden-of-Eden narrative, God's rhetorical question "Who told you that you were naked?" [Genesis

3:11] is posed on a higher level of thinking than the lesser species. The question recognizes that humans have advanced into a higher-dimensional perception of themselves. It goes together with the recognition of a "subjective reality" by which "I" am an individual. It indicates that mankind recognizes himself as different from the animals.[4] That higher faculty is one of the ways in which people exist in a higher-dimensional reality, not accessed by inferior creatures. God follows that rhetorical question with the accusation, "You have eaten, then,..." indicating that it is now clear that mankind has advanced to that elevated state of consciousness. From then on, God deals with mankind on a new, higher, and more sophisticated level.

To gain access to each additional dimension is to establish a connection that links it to the lesser dimensions we already have. For example, nothing that happens in chemistry violates the laws of physics; nothing in biology violates the laws of chemistry. Stepping upward: in higher animals, any voluntary movement initiated by the mind flows through the brain to the electrical nervous system to synapses and employs the chemistry of muscle contractions.

In every case, there is an association between the consecutive steps that connect the higher dimensions, back to the level of measurable phenomena. Still further upward: enjoying beautiful music involves a cascade of highly com-

---

[4] B. J. F. Lonergan, *Insight*, ch. 6, sec. 2.5.

plex behaviors, including discerning interwoven themes, interpreting pitch and overtones of frequencies while listening, but also reaching all the way back through audio equipment to the vibrating eardrum and the physics of sound waves.

On the level of interactions within a complex society, the ability to construct social policies based on law, politics, and history necessarily relies on yet another cascade of co-operating steps, including the characteristics of memory, speech, and judgment. The output of one mind is the input to another, and results are *real*. These higher realities are built on a lower platform, but not *reducible* to a collection of lower functions.

Still higher up, the phenomenon of human love lies on a very high plane. As everyone happily smiles at a cute little toddler, husband and wife can recognize that the child is *their love for each other* walking around. Human beings can perceive in what they create a sharing in (and reflection of) God's creative love.

The human being naturally wants to advance to additional higher stages. We strive to access the "invisible" part of God's created reality. The Christian sees this as moving in the direction of harmony with God. Teilhard de Chardin spoke in terms of *complexification* and imagined that harmony as *Christogenesis*. As illustrated in appendix C, we're moving along his outbound arrow.

The Pathway

The obvious questions arise: "How can I do this? Where is a guidebook?" The Bible serves very well. Recall that when Jesus turned water into wine at the wedding feast of Cana, that *really* got the attention of the people. But that event was on the level of atoms and molecules. When they were ready to listen, he presented the eight beatitudes as a guide to living according to the Kingdom of God. Notice that every one of the eight is at a higher level of humanity. For example, mourning and comforting involve the interaction of truly *human* characteristics. And peacemaking is the most difficult job in the world.

Jesus constantly urged people to take an upward step to a better life. When he said "Come, follow me," He didn't merely mean "Step out of your fishing boat," but "Change your life to an entirely new way of living in unity with God."

For a person reading the Bible afresh, there is hope for a new insight *today*. We seek new pathways of perception to better appreciate truths God makes available to us. But we need to be aware that our languages (crafted from sensory perception and experience) will take us only so far.

Moreover, God has given to some humans exceptional access to a level of spiritual awareness (in still higher dimensions) which they cannot express in words to those lacking comparable experience. Christian mystics over the centuries have acknowledged this obstacle to communication; and the difficulties of translation exacerbate the problem. External guidance is pretty scarce to those scaling high up the ladder.

There comes a point where the only remaining option is to accept one's limitations. No human during his/her lifetime is going to rise beyond a certain level, and that varies from person to person.

Continuity of the Higher Dimensions

Acknowledging the inevitability of bodily death invites an interesting speculation: At death, when the atoms and molecules return to dust, the higher-dimensional reality that *is* a human being loses its connection to the four dimensions of space-time. The many remaining dimensions, although uncoupled from space and time, certainly need not vanish. Their existence is independent of the platform of the body in four-dimensional space-time. Time, which was never really "absolute" in the Newtonian sense, is simply not involved. Life continues in a way unrelated to time. (I add a caution here: the word "continue" in ordinary parlance often conveys a sense of time, which isn't the case here.)

Time does not "stand still;" the clock does not "run forever." The word "eternal" indicates that time simply is *not* one of the variables.

The trouble with this picture is that it defies description because description requires words in a conventional language. This is very unsatisfying to nearly everyone; we expect a description in terms of recognizable images or analogies. But think for a moment about what Christians are taught to anticipate: When St. Paul wrote "Eye has not seen and ear has not heard..." [1 Corinthians 2:9], he was essentially saying (expressed in contemporary physics terms) that the new form of life is uncoupled from eyes and ears, from sensory perception, from measuring instruments, from electromagnetism, from space and time.

Assorted Bible verses can be interpreted to support this "extra dimensional" perspective, but they're not very persuasive. Conceptually, a person may agree that the word "after" isn't exactly right, but no one has a good alternate way of speaking. While originality and novelty have considerable appeal, any explanation that starts from physics equations will certainly be only a partial answer.

The urgent question "what happens *after* I die?" is still on everyone's mind. This has always been in the realm of mystery (the *transintelligible*). We accept that further description necessarily will be scant and limited to analogies. Long ago, Christianity acknowledged that heaven is not a *place*, but rather a *state of being*. However, so far it has been too big a step for people to say that heaven is not a *time* either. This topic will be revisited in chapter 12. Modern physics may perceive time as on the same footing as space, but people have great difficulty breaking free from their usual perception of time.

Conclusion

Human perception is very distorted. There has been a long and sad history

> **Heaven is not a place, but rather a state of being. So far it has been too big a step for people to say that Heaven is not a time either.**

of humans imposing such limitations upon God. What we have learned about space and time in the past century points to the reality of more dimensions than just space and time alone.

Materialism and *scientism* won't consider taking any step in that direction, but the Christian creed invites us to look more deeply. A "dimensional" way of framing certain questions may have merit: we can instantly accept that God manages countless dimensions, enabling His creatures to access some of them.

The human endeavor can be seen as a progression upward toward God, rising through additional higher dimensions. However, any attempt to assign a correspondence between particular human traits and particular dimensions would likely be futile. Rather, the need to defer to God's creative power is obvious.

One outcome of this kind of thinking is that those higher dimensions don't vanish when the connection to space-time is broken at death. A state of being uncoupled from both space and time is easy to acknowledge, even though human language will always fail to provide a description. In Part 3, we'll look further into this, employing concepts that are hard to express in regular language.

Part 2

God's Hand in Reality

In the first part, I focused attention on time, stressing how our customary perception of time causes an obstacle to appreciating God's creative powers. In this part, I'll describe several specific ways in which that power has given us the world (indeed, the universe) we see around us. God's transcendence makes Him exempt from our limitations; being *present* to all time enables God to attend effortlessly to what we would call "very short" or "very long" elements of time.

Chapters 7 and 8 describe what we think we know about our world in our universe, emphasizing that our time isn't God's time. Chapters 9 and 10 caution against the arrogance of believing in our own capabilities. Chapter 11 points out certain unusual features of time in our everyday life.

Throughout, the underlying theme to remember is that God created time and space, and God has given us the tools of science to understand His creation. Using science, we have discovered a lot about what a magnificent edifice it is.

Chapter 7

The Big Bang: God Started It All

Long before there were any written records, mankind looked at the night sky and wondered about what's out there and where it all came from. Staring at the sky all night long invited thinking up stories, and ancients came up with many fanciful explanations, most of them involving warring gods.

Ancient Astronomy

What can loosely be called astronomy and cosmology predates Aristotle; the various constellations, the Zodiac symbols and the art of casting horoscopes all go back that far. People could see some objects moving against a background of "fixed stars," and those wandering objects were called *planets*, and then given names associated with the mythical gods.

Amid the fanciful stories, there were some observers who discovered mathematical regularities of the motion of objects in the sky, and the science of astronomy became better organized. The Ptolemaic model of the known universe was established in Roman times and represented the best you could do via the naked human eye. Ptolemy's model remained the dominant explanation for those motions until the time of Copernicus and Galileo, when the telescope was invented. Across many centuries, scholars attempted to correlate motions of objects with various passages in the Bible. Later, those attempts would become obstacles to understanding.

Galileo's troubles have been accurately described as a *clash of cultures*. For many centuries, "truth" had always been traced to the philosophy of Aristotle, and that was the dominant culture of the day. Followers of the (relatively new) *scientific method*, including Galileo, relied upon observational data as their standard of truth. Galileo invited people to look through his telescope and see for themselves what was going on; but those in the opposite culture could not even fathom the idea of allowing observations to dictate the pathway to finding truth. Galileo had very little patience for such people, which exacerbated the conflict.

Going forward from Galileo's observation of the moons of Jupiter, it took only about fifty years to convince all educated people that the Copernican model being espoused by Galileo was correct. Unfortunately, the *clash of cultures* took place in a shorter length of time;[1] Galileo was put on trial by the Inquisition and confined to house arrest for the

---

[1] Giorgio de Santillana, *The Crime of Galileo* (New York: Time Inc. Book Division, 1962), and Jerome J. Langford, *Galileo, Science and the Church* (Ann Arbor, MI: Univ. Michigan Press, 1971).

remainder of his life. That terrible mistake has been a major source of derision of religion for four centuries.

Modern Physics

For most of human history, into the 20th century, we had no idea of separate galaxies. There were some fuzzy objects out there, but telescopes weren't good enough to resolve what they actually were. Nor did people appreciate the vast size of the universe. For example, it was not realized that the Magellanic Clouds were in fact two neighboring galaxies.

Just as measurements were getting much better, along came Einstein's general theory of relativity (GR), which united space and time and gravity. Einstein's theory was conceptually different from what people had previously assumed: *classical mechanics* (composed by Isaac Newton circa 1700) was considered perfect and infallible. [More about this conceit in Chapter 10.] Therefore, Einstein's theory arrived as quite a surprise and naturally was rejected by the prevailing major culture.

In 1919, a major prediction of Einstein's theory was verified via observations during a solar eclipse, and that greatly enhanced the credibility of general relativity.

Soon many more scientists got interested because it was possible to associate the huge amount of observed astronomical data with a theory that made sense of it all. The measurable difference in light arriving from some very distant stars (the "red shift") provided convincing evidence that the universe was expanding – and that begged for an explanation.

Einstein's equations of general relativity involved *tensor calculus*, which was unfamiliar to most scientists at the time; but a few set out to solve those equations for special conditions of the universe. Einstein himself believed that the universe was in a "steady state," hardly changing at all.

Around 1922, physicist Alexander Friedmann in Russia worked out a solution for a universe expanding from a singular starting point; unfortunately, he died soon thereafter, and his work wasn't noticed.

Working independently, Georges Lemaître, a Belgian Catholic priest, solved Einstein's equations for a universe starting at time t = 0 and expanding from a singular point to its present size. In that case, the entire physical universe was once incredibly tiny and dense. Lemaitre turned that in as his doctoral thesis jointly to Harvard and M.I.T. in 1925, and it was quickly noticed in the western scientific world. Indeed, the most important aspect of Lemaitre's accomplishment was to present a universe that *did* have a starting point – there *is* a beginning of time.

When Einstein heard of Lemaître's work, he scoffed at it; and that disdain put Lemaître into an uphill struggle. The impression of a steady unchanging universe "out there" was very strong in those days, and the thought of everything starting off at a single point was incomprehensible to most physicists. The famous astronomer Fred Hoyle sneeringly invented the phrase "Big Bang" to pile on the ridicule.

The disdain for Lemaître didn't last long. Better telescopes were built, and galaxies beyond our own Milky Way were found. By 1929, Edwin Hubble's observations permitted a calculation of how fast the universe was expanding, and it was all consistent with Lemaître's theory. Einstein himself eventually came to agree with Lemaitre – for the simple and honorable scientific reason that Lemaître's theoretical solution accounted for the data. Einstein's general theory of relativity, meanwhile, became fully accepted throughout the scientific world.

Lemaître became a key scientific advisor to Pope Pius XII. In 1951-52, a most interesting backstage drama took place,[2] which shows what real scientists think about even the best scientific theories.

Pope Pius XII saw that the Big Bang theory coincided very nicely with the narrative in chapter one of the Book of Genesis and was going to declare it to be true, a doctrine of faith. Obviously, that would have been a huge accolade for Lemaître, a permanent vindication of his theory. Instead of rejoicing at this, Lemaître himself talked the Pope out of it. Lemaître explained that NO theory in physics, however elegant or reliable, is truly final. *Every* theory can always be

---

[2] Michael Heller, *Creative Tension*, Ch. 8 (Radnor, PA: Templeton Foundation Press, 2003).

revised; *every* theory can be contradicted (and thereby destroyed) by a single experiment. Lemaître knew his history well: only a century earlier, "the ether" seemed a sure thing.

In 1964, new discoveries in radio astronomy gave further evidence that indeed the universe originated in a sudden explosion,[3] and the competing "steady state" theory was abandoned. Data triumphed over theory, and the Big Bang became the only game in town. However, with the passage of yet another half-century, recent observations have indicated that some correction may be necessary to Einstein's theory: there seems to be some additional force (customarily termed "dark energy") that causes the expansion of the universe to accelerate. In the years ahead, will general relativity or the Big Bang be corrected? Stay tuned.

Fr. Georges Lemaître had an enduring confidence that both science and religion are complementary pathways to knowledge, but scientific theories can stand or fall on their own and don't need religion to referee. Thinking that a scientific theory is "final" would be a false absolute. It is enormously to the credit of Lemaître that he dismissed fame for himself and retained his perspective on the proper relationship between science and religion.

---

[3] Arno Penzias and Robert Wilson "A Measurement of Excess Antenna Temperature at 4080 Mc/s". *Astrophysical Journal Letters* **142** (July 1965), 419–421.

A Universe with a Finite Past

From time immemorial, mankind's perception of the universe was that it's always been there, and there has always been time as well.[4] Time was considered a mathematical coordinate, a totally abstract entity, not a created aspect of the universe. God *existed within* a system of coordinates rather than being the creator of the coordinate system.

In the centuries following Newton's presentation of *classical mechanics*, the picture was that time goes on forever in both directions, past and future. Among scholars in those days, some thought that God chose a particular moment to create the universe while others thought that the universe had merely been here all along, forever in the past.

During the Enlightenment period, virtually all philosophers and writers based their thinking upon an infinite expanse of time.[5] Across those centuries, there was no motivation to examine the notion that time might be finite in extent.

The Big Bang theory changed all that. The Steady State model (popular for centuries) had time running forever in both directions, but the observed radio astronomy evidence destroyed that option. Several decades later, additional investigations by Borde, Vilenkin, and Guth showed that the

---

[4] See, for example, The *Bhagavad Vita*

[5] See, for example, D. Hume, *On Miracles.*

universe is "past-limited;"[6] they proved that, for *any* universe that is expanding, it *must have* a finite duration, a beginning in time. Our current best estimate is 13.8 billion years.

> **Lemaître explained that NO theory in physics, however elegant or reliable, is truly final.**

The importance of that scientific fact has so far escaped the attention of professors of philosophy. It has now become necessary to go back and re-read texts from the Enlightenment (Voltaire, Hume, etc.) and recognize that many things they took for granted are invalid. As one example, Nietzsche's presumption that the universe eventually comes back together and repeats itself is never going to happen.[7] Likewise, the notion of perpetual death and rebirth of the universe (actually traceable to ancient texts from India) is incorrect. There are many more examples. The world of philosophy has a big repair job waiting to be done.

To go down that path, the first step must be to acknowledge the findings of modern physics, that time is not an immutable abstract concept (it's not a Platonic *form*), but

---

[6] A. Borde & A. Vilenkin, "Eternal Inflation and the Initial Singularity," *Physical Review Letters* 72, (23 May 1994), 3305-3308; and A. Borde, A. Guth & A. Vilenkin, "Inflationary Space-Times are not Past-Complete," arXiv :gr-qc (14 Jan 2003)

[7] Frederick Nietzsche, *Thus Spoke Zarathustra.*

time is part of the fabric of the real universe. The conceptual shift from *abstract* to *real* means that time came into existence at the beginning of the universe. Ever since, the space-time continuum has been expanding, and what we perceive as gravity IS the curvature of space-time. This concept is a challenge to understand, even for physicists who can think mathematically and can geometrically handle multiple dimensions. It is doubly difficult for those with only a background in philosophy.

Every one of the enlightenment scholars is an example of that limitation, even if they were schooled in *classical mechanics*. None of them learned from St. Augustine that God created space and time together, and *that was* "the beginning."

Escaping from Imposed Limitations

Much of the literature known as *Process Theology* (beginning in the late 19th century) has God Himself evolving over time.[8] We are asked to believe that God is subject to time, just as humans are, and then to adapt our understanding of God to fit this constraint. Without paying attention to the point, Process Theologians have denied God's fundamental property of *omnipresence*. Another big mistake.

---

[8] Cf., Alfred N. Whitehead, *Science and the Modern World*, *Lowell Lectures, 1925* (Glencoe, IL: Free Press, 1967).

The well-known quip that "man made god in his own image and likeness" isn't just a wisecrack by atheists. It contains a warning about the danger inherent in our very limited human way of thinking and understanding, particularly on so fundamental a point as the *sequential* nature of our lives. In everyday life, absent observed data to the contrary, that's about all we can do. However, we must go back and look more carefully. We need to realize that we have distorted our understanding of God by artificially limiting Him to our human mode of thinking.

Looking at the magnificent beauty and symmetry of the laws of physics, and all that has proceeded from the origin, inspires a scientist to ask, "Who thought this up?" That's not "Who discovered it?" as in Newton or Einstein.

Many prominent scientists have pondered the question. The great 19th-century scientist Sir James Jeans (noted for much of thermodynamics) said very succinctly, "The universe begins to look more like a great thought than a great machine." Albert Einstein famously said, "I want to know God's thoughts. The rest are details."

God's principles or laws that we know as science didn't stop with the achievement of stars and planets. His ability to guide further creation in an elegant way has kept right on going. In the next chapter, we'll look in detail at the incredibly rich unfolding as creation continued to increase in complexity.

Chapter 8

Evolution: God's Method of Creating

If we assert that the Earth, as we know it today, took 13.8 billion years to become what it is – that is, if we do not accept that the earth was created in six 24-hour days and is 6000 years old – does that mean we reject the Bible? Does it mean faith and science are incompatible?

No, not at all!

The Bible: "Eternally True, Not Literally True"

I once attended a lecture called, "Are the first 11 chapters of Genesis true or did we descend from apes?" The phrase "descend from apes" is used derisively to ridicule the scientific viewpoint. However, the really troublesome word hidden away there is "or." Yes, "or," – the shortest word in the title. Some Christians cannot imagine any possibility of seeing **both** truth in the Bible **and** truth in the science of evolution. To some, they are mutually exclusive. That's the mistake!

The Bible is a magnificent accomplishment! It's by far the most valuable tool we have for learning, understanding, and improving our knowledge of God. It points us in the right direction, but that is not the same as getting there –

"there" being the perfect knowledge of God, which can only come in heaven.

Everything that man does is limited, including writing things down, including trying to understand the message that God **wanted** to give humankind. A representation in print of what God intended could never be perfect because God knows that humankind is not perfect.

What humans can write down doesn't have the perfection that God has. As St. Paul says in Romans 11:33, "Oh, the depth of the riches and wisdom and knowledge of God! How unsearchable are his judgments and how inscrutable his ways!"

The front end of the Bible is not a historic or scientific text. Genesis gives us a poetic representation to describe how God created the universe. As good as the Bible may be, it still suffers from the limitations of humans trying to read it. It's at the level at which we human beings can understand because it's in a **language**. The literal words on the page are the best effort that can be made by the people who received the inspiration from God – all of which is a nice try but can't come close to what God really meant. Therefore, there is no conflict in saying that something is both true and limited.

Recall the section in the book of Job, where, in answer to Job's pushiness, God asks him "Where were you when I laid the foundations of the World?" (Job 38:4) Of course, Job could not answer back … and neither can we. We need to humbly acknowledge our own limitations.

Taking the six days of creation literally is extremely narrow thinking.[1] Both atheists and creationists make the same mistake: they imagine *time* is absolute and immutable. To say that God exists *within time* is a huge mistake, because it makes God *subordinate* to time; it places a false god (i.e., time) ahead of God.

Little wonder that atheists disbelieve in such an inferior god. Bertrand Russell, about a century ago, sneered, "If your God is so powerful, why did it take him *so long* to make the world we see?" Often creationists accept that false premise – they agree that God exists within time – and to counter the atheist claim, they assert that God created everything about 6,000 years ago.

Contrasting the Opposing Sides

For centuries, most people took the Bible literally, despite St. Augustine's book (written 415 A.D.) about Genesis,[2] where he essentially said, 'Don't take it literally.' Meanwhile, Jewish scholars over the centuries have likewise cautioned against literalism. Rabbi Jonathan Sacks presents an extensive catalog of that teaching in a 20-page appendix to his

---

[1] Stanley L. Jaki, *Genesis I through the Ages* (Merrimack, NH: Thomas More Press, 1992).

[2] St. Augustine, *The Literal Meaning of Genesis*.

book on the compatibility of faith and science.[3] Since the 19th century, the age of the earth has been contested. The dispute between atheists and creationists surged in intensity right after World War I and has continued for over a century.

There are many detailed presentations given by atheists or by creationists, more than I can summarize here. The interested reader can turn to very old books on both sides; or find recent books: Richard Dawkins states the atheist case strongly;[4] but see also the contrary book by the McGraths.[5] On the creationist side, Van Doodewaard provides one example;[6] the website https://answersingenesis.org/genesis/ did-bible-authors-believe-in-a-literal-genesis/ emphasizes that all the Biblical authors read Genesis literally.

I'll stand with St. Augustine, rather than these authors. My point here is about *time*.

All of us who believe in the omnipotence of God agree that you cannot *prove* that God *didn't* create the world just as we see it today; but then, God might have created every-

---

[3] Jonathan Sacks, Appendix in *The Great Partnership* (New York: Schocken Books, 2011).

[4] Richard Dawkins, *The God Delusion* (Boston: Houghton-Mifflin, 2006).

[5] Alister E. McGrath & Joanna C. McGrath, *The Dawkins Delusion* (Westmont, IL: InterVarsity Press, 2007).

[6] William Van Doodewaard, *The Quest for the Historical Adam* (Grand Rapids, MI: Reformation Heritage Books, 2015).

thing just 10 minutes ago, including both you and me and all our memories. I'm reminded of Galileo's statement that God, who gave us the ability to think and reason, wouldn't expect us to forego the use of such abilities.

Creationists have heard the word "transcendent" regarding God, but they don't grasp the meaning of it. Nobody does. The concept that God is *present to all time* is beyond comprehension. *Omnipresent* may mean "God is everywhere," but it's harder to grasp that "God is everywhen."

For young Christians as they transition into adulthood, clinging to a literal belief in Genesis inevitably leads to conflict with professors in college; and without a strong support system, that can be a train wreck. Better to have confidence that faith and science are compatible.

The path toward atheism doesn't start with Marx, Freud, and Dawkins, but begins by setting up an opposition between faith and science. But who benefits by rejecting science? No one.

Evolution: Valid, But in Need of Improvement

Atheists say, "Look at evolution, and then don't believe in God"; but they are basically telling us not to believe in a god *who is subject to the slow passage of time*. However, it is terribly important to *distinguish* between the kind of god that fits within our limited scope, and the God that *really* exists.

> **The concept that God is present to all time is beyond comprehension.**

"Evolution" is the best answer science has developed to the question, "How was the world created?" and I (as a Christian) have no problem accepting evolution. That doesn't mean that evolution as formulated by Darwin is 100% perfect: like any solid scientific theory, it can require modifications as we learn more. In fact, the one glossed-over problem in Darwinian evolution pertains to time.[7]

The three cornerstone principles were random mutation, natural selection, and "deep time." By mid-19th century, it was known that our planet was really, really old; but an age of 4.5 billion years wasn't part of anyone's thinking. Within the framework of classical mechanics, time just ran forever in either direction. On an entirely qualitative basis, time was believed to be "deep enough" for any evolutionary process to take place. Nobody understood the time-scale for biological processes, so nobody investigated what "deep enough" must have been.

Darwin studied finches and saw how natural selection worked but presumed the variations were totally random, over unlimited time. Today, we understand the details of

---

[7] Charles Darwin, *The Origin of Species* (New York: Random House, 1998).

random mutations, and it turns out time is NOT deep enough – by a very wide margin. Mutations happen randomly along the DNA sequence, and most variants are harmful or neutral. The new DNA sequence gets passed down to the next generation, but each generation takes time. In four billion years, an amoeba wouldn't evolve into even a fish by random mutations alone. Think about the "minor" transition from *Homo Erectus* to *Homo Sapiens Sapiens* (modern man), which took about a million years or so. Even 100,000 generations are not enough for random mutations along the DNA molecule to accomplish that change.

Those numerical facts discredit the casual use of the term "deep time." Michael Behe goes into numerical detail on this.[8] In the same vein, Douglas Axe argues cogently against pure chance.[9]

If time *isn't deep enough*, then Darwin cannot be taken as the final word. But that doesn't mean the entire theory of evolution must be completely discarded. There is surely some major revision needed to Darwin's presumption of totally random change. One new pathway that is being explored is the concept of *cooperation* at the level of primitive entities. Mathematical biologist Martin Nowak starts from

---

[8] Michael J. Behe, *The Edge of Evolution* (New York: Free Press, 2007).

[9] Douglas Axe, *Undeniable: How Biology Confirms Our Intuition that Life is Designed* (New York: HarperCollins, 2017).

the mathematics of probability and develops some very original ideas that may become fruitful.[10]

Of course, none of this was even imagined in Darwin's day, and so the glib phrase "deep time" was accepted without scrutiny.

In the years after Darwin, the position known as *Scientism* arose, wherein the only real knowledge was said to be that which was gained from science, and all else is invalid. (See Chapter 10.) Scientism became a cornerstone of the atheists' argument that God either doesn't exist or is irrelevant. "Logical Positivism" became a dominant philosophy. However, that all came tumbling down with the 20th century discoveries of quantum mechanics, Gödel's theorem,[11] and the supplanting of Newton's classical mechanics by relativity. Classical mechanics still works fine in our daily lives, but the philosophical difference is that *determinism* no longer holds.

The current embodiment of evolutionary theory carries the label "neo-Darwinism." It's a combination of Darwin's 1859 treatise, enhanced by modern genetics understood via DNA. It is certainly not a complete or perfect theory, but it's better than any alternative. Evolution really does make a lot

---

[10] Martin A. Nowak, *Super Cooperators* (Glencoe, IL: Free Press, 2011).

[11] Appendix C in Stephen M. Barr, *Modern Physics and Ancient Faith* (Notre Dame, IN: Univ. Notre Dame Press, 2003).

of sense; or rather, nothing in biology makes sense without it.

"The Language of God" by Francis Collins is an outstanding explanation of evolution and how it was all originated and intended

> **Scientism became a cornerstone of the atheists' argument that God either doesn't exist or is irrelevant.**

by God, whose intelligence infinitely surpasses our own.[12] There are many other excellent books that likewise show how God used evolution.[13]

John Haught, in particular, draws the very important distinction[14] between giving an "either/or" answer to a question, or giving a "both/and" answer. He points out that questions can be answered on different levels, and such answers need not be in conflict. The enduring fight is between those demanding "either/or" answers. Haught shows that theology

---

[12] Francis S. Collins, *The Language of God* (Glencoe, IL: Free Press, 2007).

[13] Cf., Kenneth R. Miller, *Finding Darwin's God* (New York: Harper Collins, 2000); Arthur Peacocke, *Evolution: The Disguised Friend of Faith?* (Radnor, PA: Templeton Foundation Press, 2004); John F. Haught, *Deeper Than Darwin* (New York: Westview Press, 2003); and Josef Zycinski *God and Evolution* (Washington, DC: Catholic Univ. America Press, 2006).

[14] John F. Haught, *Making Sense of Evolution* (Louisville, KY: Westminster John Knox Press, 2010).

is not trying to give scientific answers to questions. He concludes optimistically with this sentence:

> "Even though Darwin himself seemed oblivious to the potential his discoveries have to stimulate theological, spiritual and ethical renewal, his theory of evolution is a great gift to Christian theology and spirituality as they seek to interpret Jesus's revolutionary understanding of God for our own age and future generations."

There is no need to reject science! Today, there is available a much more refined understanding of theology, as well as an understanding of the limitations of science. A person who advances in science enough to discern the limitations of science will not succumb to believing in either *scientism* or six-day creationism.

A Creative Way to Reconcile the Six-Day Conundrum: Schroeder

It requires serious thought and study to comprehend that questions can be answered on different levels, that "both/and" gives the more comprehensive explanation. Here is one example of someone who followed the "both/and" route: Gerald Schroeder.

Gerald Schroeder was born in America and got a PhD in physics from MIT circa 1960. He subsequently emigrated to

Israel and became a professor. Schroeder adheres to a particular branch of Orthodox Judaism that follows Nachmonides, a 14th-century Rabbi (not to be confused with the earlier Maimonides). Nachmonides taught that the earth was created in six 24-hour days.

Then Gerry Schroeder took a most remarkable step: he decided that both his physics and his religion were true. Consequently, it was up to him to discern how the earth could indeed be created in **both** six days **and** 13.8 billion years. Schroeder's faith commitment forced him to think in a far more original way than anyone else had previously.

Because he knew the theory of relativity, Schroeder was aware of the concept of moving coordinate systems, in which time passes differently simply because of the motion of the coordinate system. When the speed of relative motion is a high fraction of the speed of light, observers in different coordinate systems will measure entirely different amounts of elapsed time on their clocks. (That aspect of Einstein's theory has been verified experimentally many times over.)

Schroeder then introduced the hypothesis that the first chapter of Genesis was written from the viewpoint of someone moving with the expansion of the early universe.[15] The second and subsequent chapters were written from the viewpoint of an observer stationary on earth. With time-

---

[15] Gerald L. Schroeder, *The Science of God* (Portland, OR: Broadway Books, 1998).

dilation due to that rapid expansion, the first day for that moving observer occupies about 7 billion of our years; the second day about 3.5 billion years; the third day almost 2 billion years, etc. Comparing clocks, the stationary observer calls it 13.8 billion years, but the moving observer calls it approximately six days.

Schroeder has solved the problem for creationists.

By insisting on finding a **both/and** solution, Gerry Schroeder got science and Scripture to agree on a centuries-old dispute. He did so by invoking a principle of physics that is well-established today; relativity is not controversial at all. The unique quality that Schroeder brought to the table was requiring of himself that both his religion and his science must simultaneously be true.

The lesson for the rest of us to be drawn from Schroeder's story is that when we find an apparent conflict, we should first trust God and then demand of ourselves more study and deeper thinking.

Blending Faith and Evolution: Teilhard de Chardin

Another brilliant innovator who thought well beyond the conventional realm was Pierre Teilhard de Chardin, who carefully considered the pathway by which mankind emerged.[16] Teilhard went further and offered a vision of where mankind is going in the future: into a unity with Christ. In appendix

> **By finding a both/and solution, Schroeder got science and Scripture to agree on a centuries-old dispute.**

C, I have graphically represented certain concepts from that book.

Just as one reads the Bible with humility and trust in God, Teilhard similarly approached reading the rocks and fossils in the book of nature, confident that he would find clues to God's wisdom there.

The guiding principle of Teilhard's line of thought was that as things become more complex in their organization, new levels of reality *emerge*. The beginning of life, the growing complexity of micro-organisms, then plants, then animals, fit very nicely into Teilhard's synthesis of consecutive steps of *complexification*. Teilhard perceived that the Creator God had built in that capability even at the lowest levels. The earliest life forms already enjoy a considerable amount of complexity. Plant life developed first, and then the alternate life form of animals emerged to live in symbiosis with plants, each thriving on the other's

---

[16] P. Teilhard de Chardin, *The Human Phenomenon* (East Sussex, England: Sussex Academic Press: 1999).

waste products. We classify the increasing complexity of animal life in terms of consecutive phyla. Higher animals developed consciousness, and the further emergence of primates and hominids followed quite naturally. Eventually that consciousness advanced into the realm of self-consciousness and reflection. Every step of the way, Teilhard saw that God's wisdom guided the pathway, until we broke forth into the domain of properties that describe the soul.

Throughout his journey of discovery, Teilhard never lost confidence in God's ability to shepherd creation through evolution, even if we humans encounter great difficulties in discerning God's plan. To Teilhard, the important thing was not how we got here, but *where we are going.*[17] Via his other writing, we learn that Teilhard was able to see over the horizon, perceiving a deeper relationship with God.[18]

The Anthropic Principle

In the latter part of the 20th century, scientists noticed some remarkable features of the numerical values that are embedded in the equations of physics. A dimensionless number can be constructed by taking the ratios of certain

---

[17] P. Teilhard de Chardin, *The Future of Man* (New York: Harper & Row, 1964).

[18] P. Teilhard de Chardin, *The Divine Milieu* (New York: Fontana Books, 1960).

properties, such as the ratio of the mass of the proton to the mass of the electron. Dozens of such dimensionless

> **Teilhard never lost confidence in God's ability to shepherd creation through evolution.**

numbers can always be constructed. However, a striking feature was noticed: if certain ratios did not have very particular numerical values, the universe would be a much different place. Over time, stars would evolve differently, and it would not be possible for us to be here. Some constants of nature are very fine-tuned to make it possible for life as we know it to exist.

That surprising and exceptional constraint upon numbers observed in nature was given the name "The Anthropic Principle."[19] In Appendix D, several of those very special numbers are described; but there are even more numerical values that must be precisely tuned. Such seemingly boring molecules as H_2O, CO_2, etc., are essential to life, and their chemical properties are unique. Barrow and Tipler write "Water is actually one of the strangest substances known to science."

The collection of precisely detailed factors required for us to be here cannot possibly escape the attention of any

---

[19] J.D. Barrow and F.J. Tipler, *The Anthropic Cosmological Principle* (Oxford, England: Oxford Univ. Press, 1986).

scientist. On the other hand, mindful that a cornerstone of Darwin's theory is that all mutations are random, scientists tend to begin with the initial supposition that all characteristics of the universe ought to be random – after all, the universe is a mighty big place, and there are hundreds of billions of galaxies, each containing about a hundred billion stars.

The eminent 20th century physicist John Wheeler put it this way:

> "...is not man an unimportant bit of dust on an unimportant planet in an unimportant galaxy in an unimportant region somewhere in the vastness of space? No! ...It is not only that man is adapted to the universe. The universe is adapted to man. ...That is the central point of the anthropic principle. According to this principle, a life-giving factor lies at the center of the whole machinery and design of the world."

It is essential to recognize that these very special characteristics of our universe completely preceded anything Darwin observed or hypothesized. This remarkable evidence of intentional design is implicit in the laws of physics. Biology came along much later.

Admiring God's Creativity

Faced with the stunning numerical values that support the Anthropic Principle, one very plausible response is expressed in the hymn *How Great Thou Art*: "O Lord my God, when I, in awesome wonder, consider all the worlds Thy hands have made..." Similarly, Psalm 19 begins, "The heavens declare the glory of God."

A Bible verse that means a lot to me is from the book of Job 42:2-3. Conceding God's infinite superiority, Job humbly says:

> "I know that you are all-powerful; what you conceive, you can perform. I am the man who obscured your designs with my empty-headed words. I have been holding forth on matters I cannot understand, on marvels beyond me and my knowledge."

To stand before God as an honest scientist is to say, "I know a lot, but what I *don't know* is so much greater." To peer over the horizon of knowledge leaves me with a sense of awe at what God accomplished so cleverly – "matters I cannot understand, marvels beyond me." I could not have thought any of this up.

The astronomer Fred Hoyle (who coined the term "Big Bang") said later in life:

"A common-sense interpretation of the facts suggests that a super-intellect has monkeyed

> To stand before God as an honest scientist is to say, "I know a lot, but what I don't know is so much greater."

with physics, as well as with chemistry and biology, and that there are no blind forces worth speaking about in nature. The numbers one calculates from the facts seem to me so overwhelming as to put this conclusion almost beyond question."

A sizable number of scientists agree with this. As one expression of that perception, here is a quote from Paul Davies, a physicist who won the Templeton Prize for his books at the intersection of religion and science.[20] He sees in God's exquisitely designed *laws* of the universe a much more elegant pathway of creation than either the randomness of the atheists or the specific manipulation of the creationists.

"Traditionally, scientists assumed that the origin of life was a chemical fluke of stupendous improbability, a quirk of fate unique in the entire cosmos. If so, then we

---

[20] P. Davies, *Science and Religion in the 21st Century*, Lecture to Templeton Foundation, 2000.

are alone in an otherwise sterile universe, and the existence of life on earth in all its exuberant glory, is just a meaningless accident. On the other hand, a growing number of scientists suspect that life is written into the fundamental laws of the universe, so that it is almost bound to arise wherever earth-like conditions prevail. If they are right – if life is part of the basic fabric of reality – then we human beings are living representations of a breathtakingly ingenious cosmic scheme, a set of laws that is able to coax life from nonlife and mind from unthinking matter. How much more impressive is such a magnificent set of physical principles – which bear all the hallmarks of design – than the sporadic intervention of a Deity who simply conjures these marvels into existence."

In my view, God deserves all the credit for that "breathtakingly ingenious cosmic scheme."

There remain a finite number of scientists who simply cannot bear to recognize God's hand. Their escape hatch is to postulate the existence of a *Multiverse* – an infinite number of other universes. In that case, the universe we experience is just a random occurrence, "the one that got lucky."

The lame excuse that is the Multiverse is a totally non-scientific concept, and moreover is intellectually incoherent. The next chapter explains why the multiverse concept makes no sense and should be dismissed.

An even more thorough examination of all varieties of the Multiverse was carried out by Fr. Spitzer,[21] and when each avenue is followed to its end, the verdict is similar: "incoherent."

> **God deserves all the credit for that "breathtakingly ingenious cosmic scheme."**

Therefore, the atheists' dependence upon the *Multiverse* collapses.

Conclusion

Returning now to the "descended from apes" line: Today we know from DNA mapping that human DNA is about 98.5% the same as that of other primates. (It's some lesser percent similar to lobsters, etc.; the DNA of all living things shares a lot of very basic characteristics.) We are not smart enough to discern from DNA how God brought about the higher-intellect properties that so clearly distinguish *Homo*

---

[21] Robert J. Spitzer, S.J., & John Sinclair, "*Our Universe: Theism*" in *Theism and Atheism: Opposing Arguments in Philosophy* (Ed. Koterski & Oppy (New York: MacMillan Reference USA, 2019).

Sapiens Sapiens from lesser hominids and primates. But, as we look beyond our present very limited knowledge, we have complete confidence that God intended this outcome.[22] All the other branches on the tree of life are incidental variations that decorate God's primary plan.

There is no conflict between believing that life on earth evolved over millions of years and believing in the Bible. A loving, all-powerful God – the God described in the Bible – can create humans via the process of evolution.

God is not watching the clock the way human beings are. Time is something we as human beings are stuck with, but God isn't subordinate to time. To interpret the six days of creation in Genesis literally is to try to impose a human limitation on God. And the evidence God has set before us everywhere reminds us never to curtail our image of God.

By believing in the symmetry principles underlying the laws of nature, and following the mathematics, scientists can explain a lot and predict a lot. But every such step points further ahead, toward the brilliance of the Creator who gave us such an intelligible universe. It is astonishing to realize that in this way, God could actually create an intelligent being that is capable of loving God in return.

---

[22] John F. Haught, *The New Cosmic Story* (New Haven: Yale Univ. Press, 2017).

Chapter 9

The Mistake of the Multiverse

Some scientists give credence to the idea of the *Multiverse*, in which there are an infinite number of other universes "out there," which are totally unobservable from our own universe.

The multiverse notion is a rather recent invention. Before the discovery of the *anthropic coincidences* associated with our own universe (Appendix D), there was no particular reason to think of other universes. But we now realize that the initial conditions of our universe are incredibly fine-tuned, accurate to one part in an incomprehensibly large number. If it were not so, life as we know it could not exist, and we wouldn't be here to discuss it. Many physicists who have taken the trouble to think through the implications of this fine-tuning find that it points to our universe's having been deliberately created by a transcendent God.[1]

In opposition to such a conclusion, belief in a *multiverse* permits the assertion that our very precisely-tuned universe exists just by chance alone. That gets around the disquieting implication of God. Hypothesizing a multiverse also has the

---

[1] See, for example, Appendix 1 of *The Soul's Upward Yearning* by Robert J. Spitzer, S.J. (San Francisco: Ignatius Press, 2015).

convenient feature that it is impossible to *disprove* something guaranteed to be unobservable.

The foremost complaint against the multiverse has been that it is an obvious violation of *Ockham's Razor,* which says: always choose the simplest theory that explains the observable data. Phrased another way: don't festoon your theory with things that are unobservable-in-principle.

A generation ago, Kitty Ferguson's book gave popular explanations of various speculative physics concepts. The testing requirement that scientific theories be falsifiable was discussed. This bears upon the multiverse notion: "...a theory with no possibility of falsification isn't considered a very strong theory."[2]

When you go beyond the boundaries of science, you have stopped doing science and are thinking in some other domain. Religion is one such example, and it freely recognizes that it deals with an unobservable God – although religious people state that things which *are* observable point the way to God. For some who are antagonistic to the concept of God, *any* other explanation is preferable, and so they grasp at an alternative such as the multiverse. It generally escapes recognition that *belief* in a multiverse is a variety of religion, too.

---

[2] Kitty Ferguson, *The Fire in the Equations* (Grand Rapids, MI: Eerdmans, 1995), p. 44.

Imagining Infinity

Joseph Stalin notoriously said, "One man's death is a tragedy; a million deaths is a statistic." Stalin knew that humans are simply unable to deal with numbers beyond some point.

Believing in an infinite number of universes is made easier by *not* understanding the mathematical concept of infinity. One commonplace image is just "a number too big to count," but that's incorrect. Long ago, mathematicians gave that number the whimsical name "kiglywig," and designated it by an X with little circles at the end of all 4 branches. Another conceptual misunderstanding of infinity is typified by the expression "to infinity and beyond" by the character Buzz Lightyear in the cartoon movie *Toy Story*. You can't go beyond infinity.

In fact, you can't even get there. It just keeps receding. No matter how far you go, there is always farther to go, indeed an infinite distance farther.

In mathematics, we often deal with the concept of limits,[3] where we imagine shrinking down from a small finite increment to an *infinitesimal*, and that is used to define a finite *derivative*. Likewise, we can handle the concept of a sum over an infinite number of such infinitesimals (*integration*),

---

[3] See, e. g., F. B. Hildebrand, *Advanced Calculus for Engineers* (Upper Saddle River: Prentice-Hall, 1949).

which yields a finite number for the *integral.* However, nobody has ever actually done an infinite calculation; every calculation ever done is finite. Even the biggest computers recognize when to stop calculating and present an answer

> **You can't go beyond infinity. In fact, you can't even get there.**

which is accurate enough. It is only when *reasoning* about mathematical formulas that we freely use the symbol for infinity – it *never* enters a numerical calculation. Indeed, if a computer program is poorly written and allows registers to overflow as a computation proceeds toward infinity, it would wipe out everything else, just like a virus.

As for the real universe, that is finite. Astronomers realized long ago that if there *were* an infinite number of stars, the sky would be brightly illuminated in the daytime. The Hubble Space Telescope's famous *deep field image* displayed an uncountable *but finite* number of stars. We estimate that there are about a hundred billion (10^{11}) galaxies, each containing about 10^{11} stars, but that's still a finite number – and it's nowhere near infinity. In fact, no number, however large, is anywhere near infinity.

Pure Chance

The notion of the multiverse asserts that there are an infinite number of parallel universes (or consecutive universes), so that everything imaginable actually comes true someplace/sometime or other. With a multiverse, the very fine-tuning needed for the anthropic coincidences present in our universe are said to be the product of sheer chance alone. Countless other universes didn't have the right combination of features to enable sentient life. Ours is the one where all the special numerical ratios worked out fine. That we are here to actually experience it is just one aspect of random chance.

Just as a calibration point, keep in mind that the probability of our universe happening completely by chance is one part in $10^{(10^{123})}$, a figure known as "the Penrose number." Meanwhile, there are only around 10^{88} particles in the known universe, which is closer to $10^{(10^{2})}$, so there is a huge gulf that makes it impossible even to write down all the zeroes in $10^{(10^{123})}$. Nevertheless, these are all finite numbers.

It bears mentioning that if there *are* an infinite number of universes, then within that scope there are an infinite number of universes *just like ours*. Everything imaginable occurs in not just *some* universe, but in an infinite number of them. That's a consequence of the meaning of the mathematical term "infinite."

In his 2003 book *Modern Physics and Ancient Faith*, Stephen Barr wrote about one version of the multiverse:

"In the many-worlds interpretation, it is an inescapable fact that reality is infinitely subdivided, and that each human being exists in not one, or even a few, but in an infinite number of copies, with infinitely various life experiences. In some branches of reality you are reading this page, in other branches you may be lying on a beach somewhere, or sleeping in your bed, or dead."[4]

If you follow this line of thinking, absolutely everything you can imagine has to come true someplace, and *you* are replicated again and again, even down to the last detail of hairs on your arms. "Again and again" isn't really the right expression, because there are an infinite number of you – and an infinite number of me, too.

The Unthinkable

The catalog of anomalous universes goes on and on, far beyond my ability to type in a lifetime (which is below the mere terabyte range). There are some very ugly universes: for example, where Hitler actually wins World War II, or Stalin takes over America, or mankind wipes itself out in a nuclear Armageddon, or a nearby star blows up and destroys all life. And it's not just a few such universes; there are an infinite

---

[4] Stephen M. Barr, Op. Cit, p. 250.

number of every imaginable terrible universe. Sorry, but that's the meaning of the mathematical term "infinite."

Among those who prefer to believe in the multiverse rather than believe in God, some point to the Holocaust, and say that a loving God couldn't possibly allow such a thing, so therefore God must not exist. However, they fail to note that the conditions associated with the non-God multiverse guarantee countless Holocausts, and all the universes where Hitler wins WW2 will contain *successful* Holocausts. The task of even contemplating an infinite number of Holocausts overwhelms the mind – and hence nobody does it.

It is this overwhelming obstacle that makes the concept of the multiverse *incoherent*. No human being can tolerate its consequences. Truly zero people are willing to follow the concept of a multiverse to its logical conclusion. Even the most devout atheist must admit that a multiverse is too horrible to imagine, and emphatically *not* a suitable explanation for the universe we live in.

String Theory

Proponents of the multiverse concept would probably like to reduce their hypothesized number of universes to something finite, somewhere around the Penrose number $[10^{(10^{123})}]$. That would put our own universe within reach of probability.

The various avenues of speculative physics-theorizing that serve this purpose exploit the enormous flexibility of *String Theory.*[5]

String Theory is notorious for having very few constraints, and countless versions of String Theory are possible. Starting

> **A multiverse is emphatically *not* a suitable explanation for the universe we live in.**

from such (plausible) concepts as *vacuum energy* and *inflation*, we find about 10^{500} available choices. This enables one to develop the imaginative conjecture of the *Landscape*. Therein, it is possible for a new universe to begin by pinching off a tiny region of space-time, which undergoes a big-bang and inflates, and this process continues indefinitely. There is sort of a "froth" of new universes constantly bubbling up.[6]

It is worth mentioning that 10^{500} is still only $10^{\wedge}(10^{\wedge}2.7)$, which is a really tiny number compared to $10^{\wedge}(10^{\wedge}123)$. In the *Postscript* we read,

---

[5] S. James Gates, Jr. "Superstring Theory: the DNA of Reality" video lectures (Chantilly, VA: The Teaching Company, 2006).

[6] This is treated in much more detail within Robert J. Spitzer's *New Proofs for the Existence of God* (Grand Rapids, MI: Eerdmans, 2010) in a section called *Postscript to Part One: Inflationary Cosmology and the String Multiverse* (pp. 75-104).

"...10^{500} universes with different laws and constants *may not be enough* for anthropic explanation of the fine-tuning of the universe in which we live. ... The cumulative effect of all of these fine-tunings significantly erodes the probabilistic resources inherent in the landscape."

(Translation: "probabilistic resources" means that if you buy 7 *PowerBall* lottery tickets, but they sell 12 million tickets, don't expect to win.) Furthermore, each of those bubble universes must have its own initial conditions very finely tuned. No advantage is gained if fine-tuning is still required.

Following this narrative, one soon arrives at the concept of *eternal inflation*; that is, the process must go on forever – which requires the time dimension to be infinite both forward and backward. This is essentially the same point where Hume, Nietzsche, and other enlightenment writers wound up, although they had only Newtonian classical mechanics, not *string theory*, to aid their imaginations. We're back to infinity once more.

Continuing with the bubbles, the prominent string theorist Alexander Vilenkin put it this way:[7]

---

[7] Alex Vilenkin, *Many Worlds in One* (New York: Hill and Wang, 2006).

"...in the worldview that has emerged from eternal inflation, our Earth and our civilization are anything but unique. Instead, countless *identical* civilizations are scattered in the infinite expanse of the cosmos." Indeed, clones of each of us are endlessly reproduced throughout the inflationary universe, for "the existence of clones is ... an inevitable consequence of the theory."

And, of course, that brings with it all the unpleasantries mentioned above.

Conclusion

It finally comes back to the task of explaining our own universe, comprised of "only" about 10^{88} particles.

A scientist is always free to abandon the quest on the grounds that the answer lies beyond the boundaries of science. But for most, that's an unsatisfying place to stop. Many scientists have concluded that the evidence within our universe points to a creator who transcends the universe. That is the most reasonable and responsible conclusion to draw. Those who choose the alternate hypothesis of a multiverse haven't thought carefully about its implications and are following an intellectual path that is incoherent.

Chapter 10

The Danger of Hidden Idolatry

Among scientists who attend to topics of religion, one of the favorite quotes[1] from the 20th century is this couplet by Pope John Paul II:

> Science can purify religion from error and superstition; Religion can purify science from idolatry and false absolutes.

Very few scientists have ever thought much about the word "idolatry;" the typical reaction would be "who, me?" In science, it's hard to imagine what the word "idolatry" could possibly mean.

I once heard a remarkably concise definition of idolatry. Loosely translated from Hindu lore: confusing your own *concept* (or model or image) with the *actual reality*.

Whatever the origin, that's a pretty clear warning not to think that your own understanding of God is fully accurate. When the Ten Commandments prohibits making graven images, we immediately think of physical objects like a golden calf standing in for a god, and of course we see the

---

[1] Pope St. John Paul II, *Letter to George Coyne* (1987).

folly of that, and take the warning seriously. Several faiths disapprove of any images at all, lest those inferior representations become the object of worship.

About the only graven-image superstition still around today is the essentially humorous custom of burying a statue of some patron-saint-of-realtors in your back yard in order to make your house sell faster. That gets a chuckle out of almost everyone, including the folks who plant the statue.

Historical Example: Classical Mechanics

But now, let's think about science and examine that definition of idolatry again. For several centuries now, we have had some scientific models of nature that are exceptionally good. Newton's *classical mechanics* accounted for things in motion on earth and even for the motions of the planets. Soon slight abnormalities were identified, and a century later the French mathematician Pierre-Simon de LaPlace introduced perturbation theory to explain them.[2] By the late 19th century, classical mechanics was such an all-encompassing complete theory that scientists believed it represented nature perfectly. Everything in nature was believed to be *determined* in advance through classical mechanics.

---

[2] Pierre-Simon LaPlace, *Celestial Mechanics* (1787).

That was idolatry: thinking that your model truly represented the underlying reality perfectly. Around the turn of the 20th century, nobody thought to call it idolatry, but churchmen who accepted the prevailing scientific theory found themselves backed into a corner trying to defend the notion of free will. It was a very awkward time for religion.

Events of the 20th century exploded that particular idolatry, and today we have "The Standard Model," a combination of *quantum chromodynamics* (QCD) and *general relativity* (GR). It's an uneasy partnership, with many physicists attentive to the need to patch things together and refine the model.

A danger is arising once again of believing that a new model represents nature perfectly; that's the theme of Hawking's book "The Grand Design."[3] All physicists wish for a theory that would unite the four forces; it's colloquially known as a "theory of everything." The temptation toward false absolutes is always there. Forbearance against that temptation is a virtue owned by those who remember the history of physics.

The cornerstone principle that has saved physics is the predominance of observational data over theory. Richard Feynman's famous quote is taught to every grad student: "It doesn't matter how beautiful your theory is, it doesn't matter

---

[3]   Stephen Hawking and Leonard Mlodinow, *The Grand Design* (New York: Bantam Books, 2010).

how smart you are. If it doesn't agree with experiment, it's wrong." Today, physicists are puzzling about dark matter and dark energy because data trumps theory. The phrase "facts are stubborn

> **"It doesn't matter how beautiful your theory is. It doesn't matter how smart you are. If it doesn't agree with experiment, it's wrong."**

things" comes to mind. In a conflict between data and theory, it is theory that must change. Idolatry is not allowed.

Example from the Life Sciences

Meanwhile, over in the life sciences, the theory of evolution has great explanatory power, and is far more comprehensive than any alternative theory. The combination of Darwin's three cornerstone principles – random mutation, natural selection, and deep time – coupled with knowledge of DNA and genetics gives us the *"Neo-Darwinian Synthesis,"* which is virtually unassailable. A large majority function with the perception that this model represents nature *perfectly*. That's a false absolute. Unfortunately, this belief is so dominant that you're better off changing careers rather than challenging it.

As discussed in Chapter 8, some scientists have carefully noted the age of the planet (4.5 billion years) and calculated that "time isn't deep enough" for all the needed mutations. Others have argued that "there is irreducible complexity." The prevailing orthodoxy of biology imperiously responds that pretty soon those annoying little discrepancies will be figured out, so get out of the way. There isn't even a faint echo of Pope John Paul II's words cautioning against false absolutes.

Contemporary Example: Computer Models

A new temptation towards idolatry has arisen along with the advances in computing power in recent decades. In the 1960s, a correct computer model took us to the moon and back, which certainly boosted confidence in computer models. Unfortunately, computer models have since been pushed far beyond their capabilities.

Also taking place during the 1960s, there was a major computer model at MIT that calculated "industrial dynamics" for business forecasting. By the 1970s, that model was adapted and expanded into an enormously complex model of the *future of the planet*, known as "The Limits to Growth" model.[4] It tried to forecast quantities including food and population over a century forward. The output results were

---

[4] Donella H. Meadows, Dennis L. Meadows, Jørgen Randers, and William W. Behrens III, *The Limits to Growth* (Rome, Italy: Club of Rome, 1972).

sold as accurate representations of what *will happen* under certain conditions (determined by the input assumptions).

There was great momentum behind the model, mostly provided by people who *wanted to believe* that computational models are smarter than the programmers who wrote them. International policy decisions stemmed from *believing* in the long-range predictions of that model. Cautionary phrases like "statistical fluctuations" and "data uncertainty" went unheeded in the rush to believe in a perfect theory. Two decades later, the "Limits to Growth" computation was shown to be merely *mathematical chaos* [resulting from a "butterfly effect"][5] and was discarded – but not before the UN implemented a lot of recommendations based on it, notably population-control programs.

Believing that a computer program equals truth is a unique new variety of idolatry. The pathway to idolatry can be exposed via careful examination. Anyone who has ever written a computer program knows that the model is no better than the assumptions and input data that go into the front end.

When pundits and journalists of limited scientific qualifications read summary reports *about* what the real experts said, they distill an impression of certainty where there was none. These bubble up through the popular media and are

---

[5] See, for example, James Gleick, *Chaos, Making a New Science* (London, England: Penguin Books, 1988).

rarely questioned by the general public. Regrettably, the few sharp computer modelers who say "Stop!" can be swept aside in the rush to find certainty.

> **Believing that a computer model equals truth is a unique new variety of idolatry.**

In spring 2020, a computer model from Imperial College in London, England, was used to predict the spread of the Covid-19 pandemic.[6] Within a few months it was proven wrong, but in that interval, very restrictive and expensive policies were imposed upon entire nations.

Political and religious leaders can be drawn into the sweep as well. None of them has time to search the interior pages of a thousand-page report; they necessarily rely upon trusted advisers. But advisers are busy people, too, and for a peripheral topic like computer models, they too rely upon lesser authorities. Ultimately the "certification of truth" is traceable to very weak interpretations by non-scientists who are *incapable* of grasping what the experts originally said.

In any given year, on any given scientific topic, there is always the lurking reality that well-meaning advisers may be wrong. This is the situation we face today regarding the long-term predictions of elaborate computer models.

---

[6] Neil Ferguson, *9th report from the WHO Collaborating Centre for Infectious Disease Modelling* (March 2020).

Avoiding False Certainty

As we saw in Chapter 7, Pope Pius XII was saved from endorsing a preferred scientific theory by the intervention of Georges Lemaitre, an eminently qualified adviser if ever there was one. If only Pope Leo XIII had had such a star on hand when belief in classical mechanics was so dominant. And Pope Urban VIII really needed competent (and less self-assured) advisers in Galileo's day.

Idolatry has been a recurring stain across all of human history, and each time one form is eliminated, another pops out somewhere else. It's an enduring temptation. Within science, recurring instances of false absolutes glide in unnoticed, and are allowed to propagate unchallenged for long periods of time, until the data screams out, "There's something wrong." Thus, do beautiful theories fall – and idols shatter.

A great deal of wisdom is condensed in the dictum "Religion can purify science from idolatry and false absolutes."

Chapter 11

Exceptional Aspects of Time

In selecting the title *Everywhen*, my purpose was to always keep in mind that God is present to *all* time and doesn't need to watch time go by the way we do. The ancient Hindus would describe God by pointing to things they could see and using the consecutive slogan "Neti. Neti. Neti" – which translates as "Not that. Not that. Not that."[1] Similarly, I cannot say how God comprehends time, but there are countless examples of how His way is "not like this or that" in our customary treatment of time.

In this chapter, three distinct aspects of time that are relevant to humanity are examined; each of them has an unusual and somewhat anomalous relationship to time. Our conventional thinking about time isn't applicable.

1. Prayer and Time

Back in chapter 4, in discussing God's superior comprehension of time, it was said "God's ability to answer prayers

---

[1] Huston Smith, *The Religions of Man* (New York: Harper Perennial, 1965).

is not restricted by the time-sequence that humans are accustomed to." That needs some expansion.

Suppose you pray for a relative about to undergo open-heart surgery. Today, that's a reasonably safe procedure, but a half-century ago, it was very risky. Medical progress in the interim has changed the survival odds enormously. Where did that medical progress come from? The *scientific materialists* would say it's just chance. But an alternate explanation is that God responded to people's prayers by having a baby conceived who grows up to be an innovative and creative doctor. And here's the important point: There is no time-requirement on the timing of either the prayers or God's answer to them. That is, God can receive a prayer of 2020 and answer it by taking action in 1960. Such a capability boggles the human mind, but it "comes with the territory" when God transcends time and is merely *present* to every time.

You will say that it's more appropriate to offer a prayer of thanksgiving about the patient's recovery. Okay, fair enough, but God has the ability to sort out such a distinction. Medical history holds many examples:

During the Revolutionary war, soldiers died of smallpox at a greater rate than from bullets. George Washington decided to have his soldiers inoculated, thus suffering a mild case and being incapacitated for a while but building up immunity against a more serious smallpox attack. Years later, that approach became the principle underlying vaccination. But at the time, was it a special inspiration by God that led Washington to decide on that treatment?

> **God can receive a prayer of 2020 and answer it by taking action in 1960.**

Into the early 20th century, children died regularly from pneumonia, but then penicillin was invented, and today pneumonia is a minor disease. Who remembers to thank God for inspiring the doctor who made the discovery ...or still less, for the pharmaceutical innovators who converted it from a laboratory curiosity into a practical treatment? Similarly, it would be most unusual for a parent of a young child today to pray that the child not contract polio, but 75 years ago, that prayer was on every parent's mind.

Can anyone remember having "exploratory surgery"? That was obsoleted by the CT scan and the MRI. Did God inspire the computer programmers to figure out the algorithms to make those technologies happen? Would that be something that anybody prays for, either in petition or in thanksgiving?

When a senior couple celebrates their 50th anniversary, can any of the attending guests remember the prayers for a successful marriage offered two generations earlier by their parents and grandparents, now long gone? Not likely. But God, being present to all time, remembers perfectly well.

All this is extremely difficult for humans to comprehend – but remember, failure to understand is a human deficiency, not a limitation upon God. The simple prayer that "God's will be done" may seem feeble and inadequate at times, but it expresses a basic confidence that God has our best interests in mind.

2. Stopping Biological Time

It is possible to shut off, or stop, biological time, just by lowering the temperature of the biological entity. The notion of cryogenically freezing astronauts during very long space flights has been fashionable in science-fiction literature for years and promotes speculation about space travel.[2]

Here's some of the physics beneath biological processes: Whenever a biological entity is functioning normally, the biochemistry involves many small transfers of energy, and those proceed at a normal pace for that organism, thus establishing a rhythm or "biological time." Living systems on Earth generally maintain that rhythm across a temperature range spanning more than 100°C, which is roughly -50°C to +50°C. Outside that range, it's different. On the high side, water doesn't boil until 100°C, but other components of an organism fail. On the low side, greatly reducing the tempera-

---

[2] See, for example, Arthur C. Clarke, *2001 A Spacey Odyssey*, (London: Penguin Books, 1968).

ture slows down the rhythm or timing of biological processes. The viable human body temperature range is only between 30°C and 45°C.

Refrigerating something by cryogenic cooling in liquid nitrogen is enough to bring all known biological processes to a complete halt.[3] That means that no exchange of energy occurs, no chemical or biological processes take place. The "biological clock" stops, and "time stands still."

To the disappointment of science-fiction buffs, no full-sized person has ever been "thawed out," but speculation about it continues. Individual biological cells can be successfully frozen and thawed out. Transporting bull sperm in cryogenic containers has been done by cattle breeders for many decades.

As the process of *In-Vitro Fertilization* (IVF) is usually practiced, human sperm and eggs are combined, and after conception the extra embryos are frozen for later use. Those miniscule human beings have their biological clock shut off for the duration of the time they remain frozen, regardless of how long that might be. As a result, there are instances of biological twins born years apart. The thawing process is not always successful; but when it is, the delayed baby has not suffered any harm during those years of zero biological

---

[3] Thomas P. Sheahen, *Introduction to High Temperature Superconductivity,* Ch. 3 (New York: Plenum Press, 1994).

activity. Sadly, there are some frozen embryos whose biological parents lose interest later on and discard them.

It is worth remembering that when the first *in vitro* fertilized baby was born in 1978, the Pope immediately issued a statement welcoming the newborn into the world. He acknowledged the intrinsic worth and sanctity of the life of this human being, without in any way endorsing the procedure.

We also see the phenomenon of "snowflake babies," when a frozen embryo is implanted years later in another woman's uterus and born into an entirely different family. Those children, despite a long period with time shut off, are every bit as real as those born in an "ordinary" sequence of time. Upon meeting and interacting with such a person (detecting nothing unusual), to find out later that she vanished from time for a decade is memorable.

We have here a totally new effect breaking in upon our normal perception of time. It really *is* possible to make biological time discontinuous – for a small enough entity. Neither the science fiction writers, nor the philosophers, nor the theologians have caught up with the technology yet. We have to ask all over again, "What is the meaning of time?" In this case, we're not talking about relativity, but about actually turning off the clock of a living human being. We wish we could find one fundamental principle that would make sense of it all.

3. Timing of When Life Begins

There is an ongoing debate on the matter of abortion, which turns on the question of when human life begins. The Supreme Court in 1973 stated that they need not decide that difficult question. But that statement preceded the invention of various medical technologies, especially ultrasound. Today, the reality of early human life is evident to anyone who looks at an ultrasound picture of a baby in the womb. In fact, many kindergarteners take such a picture of themselves into school to show their classmates.

> **Neither the science fiction writers, nor the philosophers, nor the theologians have caught up with the technology yet.**

But what about the *very* early stages of life, before the ultrasound detector gets a good quality picture? Every first-year biology book says plainly that life begins at conception. But for many people, that very early life doesn't count – it's disposable. Do we know when "real" life begins? There is quite a debate going on.

Perhaps Stacy Trasancos said it best in her book *Particles of Faith* in a chapter on when human life begins.[4] After

---

[4] Stacy A. Trasancos, *Particles of Faith* (Notre Dame, IN: Ave Maria Press: 2016), 156.

observing that each gamete (sperm or egg) contains half the genetic information needed for the formation of a new individual, and explaining the process by which sperm and egg combine to become a zygote, then an embryo, she states:

> "It is agreed that fertilization is the process that marks the change from two gamete cells to a new individual organism.
>
> Although the details at the molecular level of fertilization are complex, the reasoning really is that simple. It is common sense to take the uniting of the mother's and father's gamete cells into a new organism as the beginning of a new human life. But abortion advocates notoriously introduce subterfuge by opening up settled questions for debate. Notice:
>
> No one debates when an individual spider's life begins.
>
> No one debates when an individual puppy's life begins.
>
> Without exception, the only beings subjected to such strange scrutiny about the beginning of their existence are *unwanted* human children.
>
> Just as no one debates that the emergence of a new individual organism of any other species that reproduces this way is the beginning of the organism's life, no one debates when an individual *wanted* human child's life begins either."

The key step is the combining of two distinct strands of genetic information to form a new DNA molecule. That incredibly long chain of base pairs {A, T, G, C} is a totally original and unique blueprint for a new human life. It contains not only the recognizable genetic characteristics, but the entire program for development, including on-off switches and coded instructions (subroutines?) for carrying out all the steps of growth, across many stages. Think of the eruption of baby teeth or wisdom teeth: the timing of every stage of life is controlled by the encoded instructions in that new DNA molecule.

Once that DNA molecule launches its programmed trajectory, the entire genetic unfolding – development of a zygote into an embryo, then into a baby, and ultimately into a grown-up – proceeds according to plan... unless interrupted. Our understanding of genetics and the action of the DNA molecule today is insufficient to be specific about this or that on-off switch, about this or that command. But we can very plainly see this much: It's all a continuous process, and there is **no** identifiable transition or boundary before and after which the new person becomes "real."

A written birth certificate is a modern contrivance, irrelevant in the total life process. Long, long ago, when the Hippocratic Oath was first composed to include a promise never to perform an abortion, people recognized that life before birth was entirely real and worthy of preservation. As our scientific instruments have gotten better and better, able to discern smaller and earlier stages of life, they always reconfirm that reality. The search for some transition point fails simply because there is no transition point.

As the elephant Horton in a Dr. Seuss book said so eloquently, "A Person's a Person No Matter How Small."[5]

These three separate instances combine to reinforce a point about our human understanding of time: That perception is very limited and not at all the "sure thing" we have routinely accepted. Beyond simply being the fourth dimension, we have a lot more to learn about time.

---

[5] Dr. Seuss, *Horton Hears a Who* (New York: Random House, 1954).

Part 3

Possibilities

In this Section, I make an excursion into the realm of possibilities, imagining new ways of thinking about topics beyond the normal experiences of life. There is a lot of room for speculation in that domain because nobody has any historical or experimental data to provide a comparison. There are a wide range of possible quotations from Scripture that writers have used for centuries to provide some guidance. However, none of them involves extra dimensions or different coordinate systems or unusual ways to relate time and space.

While freely engaging in speculation, I present examples of innovative ways of thinking that may or may not correspond to reality – I'm unable to prove anything asserted here. Rather than trying to **convince** you to accept my pictures, my hope is that you will react by escalating your **own** thinking to a higher plane. If I've motivated you to imagine entirely new realms and explore them, then these chapters fulfil their purpose.

Chapter 12

Eternity is More Than "After"-Life

This chapter is intended to remove "science" as an imaginary obstacle to trusting that eternity is really pretty special. I assert that the word "eternal" has a meaning different from the time-limited interpretation people usually assign to it.

People who are overawed by the power of modern science often succumb to the belief that if science can't give an account of something, it must not be "real." That's short-sighted, a conceptual blunder. It doesn't give God enough credit.

Few people have studied relativity, which is barely a century old. Moreover, our everyday experience never requires us to treat time as a dimension, so there is little incentive (and considerable impediment) in the learning process. It's a "hard sell" to tell people that relativity should matter to them. Someday in the future, when such knowledge becomes standard, it will be easier to make the case.

Does it take any special gift to learn that time is the fourth dimension? Not really – the mathematics is just algebra. But here is the really big step: it requires a willingness to accept that there is "more to reality than meets the eye." A religious person says "of course" to that, but to agree to treat the dimensions of space and time on an equal footing is a real mind-boggler.

This is where the glib phrase "trust us" from physicists becomes a stumbling block. To a physicist familiar with Lorentz transformations, the unity of space-time makes perfectly good sense: the mathematics is so elegant and simple that the world "just couldn't be any other way." Such a commitment is rooted in symmetry principles, which means the *belief* that God used symmetry principles in creating everything. That's a very satisfying way to put all the pieces together and agree that God created the universe in a very elegant way.

However, it comes with one very heavy piece of baggage: If the symmetry of space and time is true, then why is time so obviously different from space? And *there* is the really hard sell: a beautiful, elegant and comprehensive theoretical understanding (creation is so God-like!) versus a person's everyday experience. When Plato tried to tell people that life as they experienced it was only their *limited perception* of a higher-dimensional reality, he confronted the same obstacle.

Expanding Dimensions vs. Contracting Dimensions

Once you've accepted time as the fourth dimension, the next step isn't hard: There are lots more dimensions than the four of space and time. As stated in Chapter 5, when doing physics, it is often convenient to transform into a *phase space*, a mathematics of many dimensions. It is an everyday procedure to treat a problem mathematically in a space of infinite dimensions. In physics research, we build grand theories and fabulous machines based on mathematics that freely uses other dimensions.

Anybody even slightly religious would immediately agree that if humans can deal with unlimited dimensions, so can God. It's a very small step further to say that God created all those dimensions. The question immediately arises, what does God do with all of them? Who knows? Angels? There could easily be other universes, other modes of existence, nothing at all like the one we know.

What we experience here is very finite. Science (physics, chemistry, etc.) is limited to the four dimensions of space and time, and every instrument can only measure the space-time properties of things. Our senses give us data exclusively from the realm of space and time. Some people take this to mean that there *is* nothing other than atoms in space and time. That viewpoint is called *Scientific Materialism.*

Scientific Materialists assume that advancing science will someday allow everything unexplained to be merged into a comprehensive theory. Next, they pretend we're already there. To denigrate religion, they begin by ridiculing the Indian rain dances to the Great Spirit as "ancient superstition," since we can now do impressive computations to forecast weather. From this, they generalize their ridicule to all forms of religious or super-natural thought.[1] They will point out that you can't prove this or that because you can't give them a demonstration *that meets the criteria they set up*: namely, some sort of measurement in space and time. Because you cannot trammel some idea into their limited scope, they deny its existence. The Scientific Materialists have concealed their limitations even from themselves.

> **If humans can deal with unlimited dimensions, so can God.**

New Images of the Soul

An enduring question throughout the ages is "What is a human soul?" Millennia ago, the Greeks thought of a two-way split between body and soul. This eventually got translated into Christian theology and remained in place well into the 20th century. The cartoon drawings of little ghosts that hang around the body is about the best way anybody could visualize this idea. Our dissatisfaction with this picture is rooted in our rejection of an oversimplified bifurcation into two distinct categories (*body* and *soul*) that are un-

---

[1] Stephen Hawking & Leonard Mlodinow, *The Grand Design* (New York: Bantam Books, 2010).

realistically separated from each other. It's fair to say that we've outgrown the old Greek dichotomy.

There is emerging a new notion of the soul that is downright reassuring. As we learn more about biology and the brain, we are coming to the realization that the human mind is something else. To be sure, it communicates with others through the senses, which are hooked into the brain, but the mind is now seen as something much more and much different from merely a gigantic collection of brain cells.

Just as a computer is made of hardware but functions via software, the brain and mind are sometimes described as having a similar relationship. If your drives, monitor, and power supply all conk out, your hardware is useless and "the computer is down," but the software exists independently of a particular machine. This computer analogy to the mind and brain is far from accurate, but it opens a door to a new way to think about such matters.

It is plausible to expect that God will reveal better ways to describe the soul to generations of the future, who function on a higher level than we do and who are better-equipped with language and experience. Today, not many scientists think that everything in the mind can be reduced to brain-cell action.

Examples of Constrained Thinking

As stated throughout this book, a deeply ingrained perception is that time is irreversible. You can go from one *place* to another and back again, but *time* only flows one way; you can't go back. Yet the symmetry underlying the laws of physics allows time to be reversible; mathematically, time is symmetric with space. However, nobody can accommodate this; in everyday life, we experience time as immutable. Yet it is easy to find examples where "time stands still." For example, if I walk away from this word processor, the machine doesn't care when I get back; there is no passage of time for an inanimate object. And the software exists completely independently of time, even if I turn off the hardware.

It is not surprising that authors fail to see their own assumption about time. Nearly all books explaining scientific advances do so. As one example, in *God and the New Physics*,[2] Paul Davies describes the advances in cosmology during the 20th century. Davies' book strives to dismantle old ways of thinking and finds religious notions absent from the world of physics. The very fundamental assumption that time is an ultimate, unchanging reality goes unrecognized. He discusses the Big Bang, the eventual collapse of the universe, and densities of matter that crush space out of

---

[2]   Paul Davies, *God and the New Physics* (New York: Touchstone Books, 1984).

existence in black holes. Throughout it all, time remains supreme and immutable. That's a mistake. Although physicists often extrapolate equations beyond the range in which they've been physically verified, it is a defective twist of physics to ascribe monumental changes to space while leaving time untouched.

When something exists outside the framework of time, time-dependent sentences impede understanding. Mathematics may still work fine under such circumstances, but ordinary language boggles. There are concepts that can be described mathematically for which words totally fail.

As another example, most theologians will tell you that heaven isn't a place; it's a new way of living. Still, they want you to bring your watch – they speak of heaven as part of a continuous forward march of time, a way of living that takes place *after* regular earthly life.

If *no place* must have *no time,* too, then most people can only visualize time standing still, as some kind of "hold that pose" freeze that lasts and lasts while time ticks on. That's an extremely unsatisfactory image of heaven; we'd rather think of cherubs sitting on clouds strumming harps. Either way, the mistake comes in trying to force heaven (whatever God really has in mind) into a mental framework in which time passes just like it did on earth.

Other Existence, Other Language

It's easy to see why parables are such a good teaching tool. All the conventional human analogies fail. The notion of time really gets in the way.

I'm willing to jump in with both feet and accept a heaven without time – a truly **other** form of existence. I'm the only Christian I know who questions an "after"-life. My issue is about the "after" part, not the "life" part.

We need to give God credit for being a lot smarter than us. Heaven is not a time-dependent entity, any more than it is a specific place. Time isn't frozen; time isn't continuous, boring, and repetitive; time simply *isn't* an issue at all. It's not one of the dimensions, not part of the heavenly existence. *Eternal* life is simply life in other dimensions, where *time is not a factor.*

If you ask me "What happens there?" I am unable to answer, because the question has an assumption about time hidden implicitly in the word "happens." Every sentence must have a verb; verbs are action words; and action denotes change with time. It's almost impossible to hold a conversation without thinking of time. Our senses have built up an entire structure around us that impedes communication about other forms of existence.

Just as on Pentecost the various listeners heard the Apostles speaking in their own languages, similarly today it is possible for those with an understanding of mathematical language to read certain words from the Bible in a unique way. St. Paul's phrase, "We see now imperfectly, as in a mirror; but then we shall see face to face" [1 Corinthians 13:12] contrasts the limits of our present life to our expectations for eternal

> **Eternal life is simply life in other dimensions, where time is not a factor.**

life. Everyone would agree to that. But the physicist detects (in the analogy of a mirror) an inference of lower dimensions in our *temporal* life than is possible in our *full* life.

There is a central point on which we can all agree. Since the earliest days, Christianity has taught that "Life is changed, not taken away." It takes the commitment of religious faith to accept this teaching because all we see with our senses is that at death, the life we have known so far is taken away.

In recent years, Fr. Robert J. Spitzer, S.J., has written brilliantly about how science points very persuasively toward God as the Creator.[3] After reviewing much convincing evidence from science about Jesus Christ,[4] Spitzer arrives at

---

[3] Fr. Robert J. Spitzer, S.J., *The Soul's Upward Yearning* (San Francisco: Ignatius Press, 2015).

[4] Fr. Robert J. Spitzer, S.J., *God So Loved the World* (San Francisco: Ignatius Press: 2016).

the necessity for only "a *little* leap of faith" to accept Christianity.

One benefit of studying a field in depth is that you find out the limits of your field and can recognize when you're inside or outside those boundaries. We are definitely beyond the limits of physics here. Having seen science from the inside, close up, scientists know how limited we are. There are some who refuse to take any further step. Some of us actually find it easier to make the "leap of faith" because at least we know where we're leaping *from*.

Symmetry principles and mathematics don't guarantee the "right answer." At best, they offer a new alternate path toward expanding horizons of thought. More dimensions, phase space, etc., are one language to help express a view of a greater reality. But knowledge of that greater reality does not originate within science. Rather, it starts with faith, hope, and love, as taught by Jesus Christ. Later, the elegant math and physics help explain it to scientists.

Chapter 13

Can Hell Be Real?

To this question, religious believers will immediately answer "yes" and are attentive to the matter of how to avoid Hell. That decision is way above my pay grade. A wonderfully insightful treatise about the state between Hell and Heaven was written by C. S. Lewis in *The Great Divorce*.[1] A major feature of Lewis' description was how surprised the people were at what they found out.

Nonbelievers say there is no such thing as Hell and ridicule the cartoons of little red guys with pointy tails and pitchforks. As Saul Alinsky observed half a century ago,[2] ridicule is an extremely powerful tool for winning an argument by denigrating an opponent. Setting up a strawman image and then ridiculing it achieves the Devil's goal of making people think he doesn't exist. This too is nicely treated by C.S. Lewis in *The Screwtape Letters*.[3]

---

[1] C.S. Lewis, *The Great Divorce* (New York: HarperCollins, 1946, 2001).

[2] Saul Alinsky, *Rules for Radicals* (Vancouver, WA: Vintage Books, 1971).

[3] C.S. Lewis, *The Screwtape Letters* (New York: HarperCollins: 1942, 1996).

The standard argument against Hell asserts that it cannot possibly last forever because a dead body obviously decays away to dust in a finite time span. That's the kind of objection I address in this chapter. I think there is a "both/ and" solution, rather than an "either/or" choice.

> **Some see in death only the decaying body.**

The basic task here is to inquire about what total *separation from God* is like. Here I propose a description of Hell which comports with the scientific account of death, decay, and deterioration; and yet matches the characteristics of Hell in the traditional Judeo-Christian understanding. Perhaps this model will have value for two kinds of people: A) those believers who want to have a more scientific inter-pretation than what is commonly portrayed in literature,[4] and B) those who adhere to the cartoon image, so easily dismissed.

The conventional scientific viewpoint is very empiri-cally-minded, attending to the realm of atoms and mole-cules, space and time, simply as science experiences it. Some scientists impose this rule as a fixed boundary around their perception of the human person, allowing no consideration of higher qualities. Consequently, some see in death only the decaying body and draw the hasty conclusion that there is no

---

[4] Cf., *Dante Alighieri's Inferno*, transl. by Daniel Fitzpatrick (St. Louis: En Route Books & Media, 2020).

"eternal life." That individual needs to re-examine the very limited set of facts and premises that led to that view.

Biological Time

First, an important reminder: All this is in the category known as *theological speculation,* which implies the old railroad-timetable caution: "subject to change without notice." At best, this can be termed "exploratory thinking." In the years ahead, as we learn more about the mind-body connection, a *much* more sophisticated understanding of death is sure to arise from new discoveries in biological science.

To pursue this, we must first appreciate the different ways that time is perceived in different reference frames (or coordinate systems). The phenomenon of *time dilation,* a well-known aspect of relativity theory that I will show here, applies to different observers in moving coordinate systems.

In chapter 11, I explained that biological time can be absolutely stopped by cryogenic freezing. Clearly, this demonstrates that biological time is not necessarily the same as time on a clock. For living things, including humans, a biological rhythm regulates the relationship between ordinary time and the "pace" of life. There is no assurance that such a relationship will still hold when a human is no longer connected to time and space.

Bodily Death

What happens at death? The body ceases to be relevant to what we commonly term "the soul" – that collection of higher functions that make up a truly *human* being. The multiple higher dimensions, which are *orthogonal* to time and space, keep right on existing, uncoupled from and independent of space-time.

The body starts to decay. The central characteristic of this process is that the information-handling ability of the brain slows down drastically and halts. Everything in our ordinary biological life experience is keyed to a standard relationship with time, and that deteriorates. Indeed, it shuts off entirely, and the body ultimately returns to dust.

External observers see this taking place on a time scale in what is scientifically termed the "laboratory reference frame." The elapsed time may seem quite short by our clock; the flat EEG in the hospital room may appear very quickly on the monitor. A fatal heart attack or stroke produces "brain death" very rapidly, and the brain stops giving off EEG signals about 4 to 6 minutes after the supply of oxygen ceases. Sometimes, other bodily functions continue, despite the apparent total disconnection of the brain from the outside world. Hence the people in *Persistent Vegetative State* or irreversible coma raise the difficult question of whether they are dead or not.

Another Reference Frame

It's possible to ask what death looks like in the "reference frame" of the one to whom it is happening. The slowing down of the brain's ability to perceive inputs, to process information, will create a backlog of yet-to-be-processed information waiting in line for neurons and synapses to function. However, these functions are grinding to a halt, and their processes only get slower and slower, asymptotically approaching zero. (Asymptotic behavior of a function is explained in Appendix A-3.) As the information-processing capability fades away, the effective time scale of the brain will become elongated, and the perception of the passage of time will thus be stretched out indefinitely.

Here it is helpful to employ an analogy dealing with the phenomenon of *time dilation*, which is part of the *special theory of relativity*. With time dilation, travelers in two different moving coordinate systems do not measure the same passage of time. Relativity explains how those differently-moving observers can draw different conclusions.[5]

This premise has been applied to spaceships moving at very high velocities. Someone aboard an interstellar spaceship rocketing away at nearly the speed of light will interpret events back on earth as taking a very long time to occur.

[5] G. Gamow, *Matter, Earth and Sky*, 2nd Ed. (Upper Saddle River, NJ: Prentice-Hall: 1965), pp 198-200.

That's *time dilation*. The apparent slowness is because the information didn't reach the spaceship promptly, but only traveled to the receding space ship at the speed of light, just a little faster than the space ship's own velocity. Meanwhile, the events back on earth took place in normal time, regardless of what the space traveler might perceive.

My new hypothesis is that the slowing down of consciousness is the biological-time equivalent of time dilation. The stretching out of the time dimension makes the deterioration of functionality "last forever" in the brain's own time frame, even though the external observer sees it all happen in a finite number of seconds. It's analogous to the separating spaceships.

Alternate Pathways

What becomes of the soul, those enduring *human* dimensions that continue existing, uncoupled from time and space?

Ideally, a transformation occurs into a new form of life, a new "state," a new relationship with God that is no longer dependent on time and space. For that person, all the bodily deterioration becomes irrelevant. In that new state, the individual interacts with God (and possibly with others) in a new way. It is optional for God to permit continuing access or participation in the time dimension.

Alternately, it is possible to make the opposite choice: Recalling *The Great Divorce,* C.S. Lewis wrote: "There are only two kinds of people in the end: those who say to God 'Thy will be done,' and those to whom God says at the end, 'Thy will be done.' All that are in Hell choose it. Without that self-choice there could be no Hell."[6]

Perhaps such a choice entails acting as though the world of atoms and molecules, space and time, is "all there is," "everything;" and thinking that God is either absent, irrelevant, or nonexistent. For that case, I hypothesize that Hell is the condition of remaining totally committed and anchored within that framework, with **no** escape from it.

We all like to speculate on who populates Hell – Dante's *Inferno* has been a classic of literature for centuries. Today, most people's list begins with Stalin, Hitler, and then expands into something akin to the Lord High Executioner's list in *The Mikado* by Gilbert & Sullivan.[7]

Duration

Hell is defined as the absence of God, God's love, and the communion of Love. I visualize it as the full, lingering experience of cessation of being, permanently and irretrievably.

---

[6] C.S. Lewis, *The Great Divorce*, p. 75.

[7] Alan Jefferson, *The Complete Gilbert & Sullivan Opera Guide* (New York: Facts on File Books, 1984).

Imagine that, based on your choices while living, at death the multiple higher dimensions were to crumble and fade away, leaving just atoms and molecules in time and space. If you watch your consciousness go away and experience the dismantling of all higher functions, including thought and feeling, that would be a horrible experience.

Because of the unique way human memory enables us to experience biological time, Hell seems to "last forever." Unlike the escape from the constraints and inexorability of time that heaven provides, it is the sequential realization of the natural biological decay process that occurs in a domain where time is supreme.

A plausible reading of Scripture says that at death a new kind of life begins. Thinking in terms of additional dimensions, this new life may well be unrelated to the dimensions of space and time. Those who, by their actions, have explicitly followed the opposite path, have chosen to permanently lock themselves into physical time and space. Accordingly, they get to experience the ultimate that space and time have to offer: death, including the biological time dilation that makes disintegration into nothingness last *forever.*

Hellfire

The authors of Scripture always spoke in terms of Hell as "fire." It's important to remember that the ancient writers were constrained by their milieu to communicate what they had to say in terms their audience could grasp. I only note that the process described here is one in which oxidation takes place, and fire is one form of oxidation. Perhaps the awareness of the oxidation of the brain, when the time frame is greatly elongated, is somehow like the perception of burning. Perhaps since burning seems a particularly slow and painful way to die (to those of us watching from the "laboratory reference frame"), the mention of "fire" was the Biblical writers' best way to convey "slow and painful."

The notion of being aware of, and participating fully in, the total decay and loss of one's consciousness is bad enough, but Christianity teaches that those in Hell are *aware* of their separation from God. So it must be that one aspect of Hell is the realization that life didn't *have* to end this way, that there *was* an alternative, now closed off forever.

Cautions

We must never confuse "being mistaken" with "being evil." It would be silly to suggest that those who regard space and time as immutable are headed for Hell. Prior to Einstein, time was held to be absolute, and for day-to-day purposes, that still works even today. Lack of scientific knowledge certainly doesn't impede sharing in the love of God. Surely, there are lots of souls in heaven who showed up there with the expectation of sitting on a cloud strumming a harp. Most people are content to await some enormous surprise.

> **Christianity teaches that those in Hell are aware of their separation from God.**

Our scientific knowledge carries us only so far, and when we try to look over the horizon beyond today's science, we should not assume that everything imaginable is going to be covered by tomorrow's science. If it were, that would be terribly disappointing.

Summary

The point of this chapter is to overcome the "cartoon" image of little red guys with pointy tails, which is so easily ridiculed, and thus has led so many people to be dismissive of the reality of Hell.

The new scientific notion that I utilize here is that of *time dilation*, which provides an explanation of how different observers can perceive the same event spanning short or long periods of time. In this picture, nobody gets bodily death over with quickly; the "way out" is to transform to an entirely new kind of life.

Furthermore, this model is silent on the terribly important question about what measure of good or evil decides if that transformation takes place, or if the interminable decay is fully experienced. The most reliable guidance we have about that is the words of Jesus himself (e.g., Matthew 25:31-46). Modern speculative descriptions (such as C.S. Lewis', or this one) carry no warranty.

This version of Hell has some similarities to the Hell familiar from Scripture. It also has some differences; one limiting factor is characteristics of language in olden days, when the concept of different perceptions of time didn't exist. Other differences are due to contemporary factors (including my limited imagination). Either way, I think most religious people would agree with the general conclusion that heaven is beyond our imagination while Hell is just what we ought naturally to expect when separated from God.

Through free will, God constantly offers people the opportunity to make choices, good or evil. God presents lots of options. The choice is possible to reject God and remain entirely confined to the limited world we know. When I pursue the logical conclusion of a living system confined entirely within the boundaries of space and time, including the termination of molecular life, I give the name "Hell" to what I see.

Chapter 14

Our Multidimensional Life Beyond Time

When we think about death – our own or a loved one's – we want so badly to cling to space and time – to "meet" a loved one "again" in Heaven. Space and time are the only experience we have to hang our identity on. But what if, freed from the constraints of the body, the constraints of space-time, we are granted access to marvelous other dimensions of great beauty and joy?

How would that work? Here I'll give you my best guess. It requires an innovative new concept regarding time.

Today, thanks to the theory of relativity, we know that time is a dimension. There is a dimensional symmetry in the way we speak of "points in a space-time continuum." Each event, each moment, occurs at one point in space-time. Relativity has opened a door to greater insight about time.

For most people, under nearly all circumstances, it is a harmless oversight not to see time as a dimension. Since chemistry and biology do not *need* to treat time as a dimension, it is very easy to forget that time *is* a dimension. That is the state nearly all of humanity is in today.

A Second Look at Flatland

To make progress, first let's step downward in dimensions, returning to *Flatland* (Chapter 5 and Appendix B). The analogy helps us learn how to look outward toward higher dimensions. When Abbott wrote *Flatland*, time just marched on, not exceptional to either flatlanders or the sphere. Now let's pay close attention to time as one of the coordinates in Flatland.

Start by applying the "dimensional" notation to Flatland.[1] By coming home every night, Mr. A. Square could occupy the same coordinates {x_0, y_0} at different times t_1, t_2, etc.; each such time constitutes a distinct coordinate in his three-dimensional *space-time continuum* {x, y, iCt}.

The sphere visiting Flatland had access to all points {x, y} because he had that extra dimension {z} in which to move. Because the sphere could vary his z-coordinate, he could appear and disappear from Flatland at will.

Now think about this possibility: Consider what kind of "magic" the sphere might have done if his *time* coordinate also were available for excursions of the sort he could per-

---

[1] The way to designate a point is space is by specifying its coordinates {x, y, z}. To designate a point in space-time, the coordinates are {x, y, z, iCt}, where t is time, C is the speed of light, and i is the square root of (-1), called the "imaginary number." ["Imaginary" is a poor word choice – it really exists.]

form with the spatial dimensions. Since the sphere has a dimension to spare in his dealings with Flatlanders, he might elect to pick a completely different *time* coordinate. If the sphere should choose a much smaller numerical time coordinate, our customary word for that is "earlier": the sphere might come to Flatland when A. Square is just a child. Likewise, he could choose a much larger numerical time coordinate; that is, "later" – and revisit Flatland when A. Square's grandchild is all grown up, with only distant memories of what his grandfather tried to teach him. Each such time would constitute another story about Flatland.

Because he is a higher-dimensional being, the sphere has a measure of *control* over Flatland that the Flatlanders themselves do not have. Keep that in mind as we return to our realm and imagine accessing God's expanded reality.

Expanding Our Perception

Taking this lesson further, we begin to ask questions about what may be "out there" amid the higher dimensions. That's an excursion beyond science, beyond space-time, beyond any means of experimentally rejecting or confirming a hypothesis. It's okay to go beyond the borders of science, but we must recognize and acknowledge the border that has been crossed.

Everyone would like to achieve genuine cognizance of realities above and beyond our present state, but how? Like

A. Square wanting to follow the sphere's lead, we're bewildered

> **It's okay to go beyond the borders of science, but we must recognize and acknowledge the border that has been crossed.**

by all the unfamiliar new concepts. It's easy to acknowledge that extra dimensions exist but trying to explain anything about them involves reducing them to ordinary language, and that requirement is a major constraint on communication.

It is mind-boggling enough to deal with time as if it can be treated like a dimension, and there are plenty of people who can go no further. It is a leap of faith-in-physics to say that time is a dimension. To allow the possibility of still higher dimensions is to acknowledge the freedom and power of God. To accept that what we express by the word "dimension" is really something much deeper is to yield to the wisdom of God.

On one hand, mankind's thoughts and insights can expand indefinitely, but on the other, it's hard to criticize people who don't advance in that direction. Many people choose instead to live a life guided by that small fraction of knowledge that God has revealed to mankind.

Extra Time-like Dimensions

The sphere could go in and out of the plane {any choice of x_1, x_2} because he had that extra dimension (x_3) to move around in. Now we want to apply that notion to our own situation, imagining a being having access to more dimensions. With more spatial dimensions {x_1, x_2, x_3, x_4, x_5, ...} that multi-dimensional being could go in and out of conventional human existence (like the world) because he could choose to occupy whatever coordinates {x_1, x_2, x_3} he likes.

To go in and out of the *time* dimension iCt is conceptually more difficult because our normal way of thinking does not treat iCt as a spatial dimension. We think of time quite differently, not as a spatial dimension at all. But should God be bound by so restrictive a notion of time? Hardly! Again, we're confronting a human limitation.

To better grasp the idea of moving in and out of the time dimension, imagine a multi-dimensional being who exists in a space {x, y, z, iCt, u, v, w, ...}, where one or more of the additional dimensions has a property that makes it temporal, or time-like. That is, just as x, y, and z are interchangeable, let us suppose that one of the extra dimensions is interchangeable with time t. Then our multi-dimensional being has an extra temporal dimension at his disposal, and thus he can move freely between the two different temporal dimensions. He can go "visit earth" at times of his own choosing by moving around in the alternate temporal variable. This is akin to the way the sphere has an extra spatial dimension to allow him to enter and leave Flatland at will.

But what *is* this alternate temporal dimension, this extra dimension that is interchangeable with iCt? We do not know. It is as difficult for us to grasp the notion of an alternate temporal dimension as it was for Flatlanders to grasp "upward, but not northward."

Because we can plainly see that the three dimensions {x, y, z} are readily interchangeable, we really ought to be willing to accept that God could create a time-domain having additional dimensional components of a temporal *vector*. It is our human limitations that cause us to think of time as a simple *scalar* quantity. As humans, we speak of the "arrow of time" – it only goes in one direction.

It is via our extra dimension of *memory* that we realize there are times other than the present. In Flatland, we laughed at the king of *Lineland*, who could see only one-dimensionally. Isn't it likely that higher-dimensional creatures scoff at us for thinking that time is immutable and one-dimensional?

As soon as one grants that God is free to make time itself multi-dimensional or (stated another way) to create a multi-dimensional universe that has more than one temporal (time-equivalent) dimension in it, then we must conclude that a multi-dimensional being has freedom of access to the many various points along any of the temporal dimensions.

Therefore, the multi-dimensional being can show up at various "times" with ease.

Of course, as humans we *do* have the experience of three spatial dimensions, but we have *no* experience of more than one temporal dimension. As a result, there is nothing in human language to convey the notion of extra temporal dimensions. Words fail, and we can't describe it to anyone. Terms like *reincarnation* and *deja vu* hint at the frustration built into any attempt to force-fit multi-dimensional existence into a single time-variable. It leads to confusion and seeming contradictions, much like what A. Square experienced when he first met the sphere before he understood higher dimensionality.

Higher Dimensions Bring Greater Freedom

Imagine that you were a being of dimensionality higher than normal space-time, that is, a being who exists in $\{x, y, z, iCt, u, v, w, ...\}$. With many dimensions available, any subset is accessible. To a higher-dimensional being, any of these dimensions may be freely traversed, going back and forth at will.

Even time? YES! iCt is just one of many dimensions a higher-dimensional being is free to traverse. For you to appear at point $\{x_0, y_0, z_0, iCt_0\}$ and then select different coordinates, including earlier or later times, is no harder than the sphere popping in and out of Flatland. You do not "travel backward" through time. Like the sphere, you merely pick one set of {x, y, z, iCt} and project your multidimensional self into it. Like the sphere, you can stay as long as you like at one spatial coordinate

Higher dimensions bring greater freedom.

set {x, y, z}, or you can move around as you see fit.

A multi-dimensional being is free to pick an ensemble {x, y, z, iCt} from his "catalog" of many dimensions and "visit those coordinates" at will. A different choice may be at a later or earlier time; iCt is just one variable. The multi-dimensional being *also* gets to choose other values for other dimensions at his disposal. We can't say much about those other dimensions since we don't know what they are. Words like "earlier" and "later," "here" and "there," only apply to the dimensions we have experienced. Again, however, there is no reason to think that a multi-dimensional being cannot be endowed by God with many additional characteristics that are beyond human comprehension.

Note that all this movement is not the same as *omnipresence*, which belongs only to God. You don't have access to *all* times and *all* spatial points, but you can choose any *one* space and time coordinate set, and then another, and another …

Going Places

One interesting question: At physical death, suppose God grants this multi-dimensionality to human souls. If so, are the space-time coordinates included in the set of accessible dimensions? We very much want that to be the case because we want so badly to cling to space and time – it's the only experience on which we have to hang our identity. But for all we know, God may grant the soul (freed from the constraint of the body, the constraints of space-time) access to brand-new other beautiful dimensions. In that case, it might not be so important to have access to space and time.

But suppose it *is* important to retain that connection to space and time. As a multi-dimensional being, you're free to be "present" at whatever points $\{x_4, y_4, z_4, iCt_4\}$ you choose. If, before you became a multi-dimensional being, you used to be a human, then you experienced a certain finite collection of points $\boxtimes\{x_j, y_j, z_j, iCt_j\}$ during your lifetime. You can have any one of them back, in its totality, complete with your full coterie of human properties, but now enhanced by all the extra dimensions you have picked up. One term for the combination of your multi-dimensional existence with the original four dimensions might be *glorified body*.

When you select particular space-time coordinates, you indeed experience the original event in its original reality. You are not "re-visiting" anything, or re-tracing steps. You are not merely watching others do something, as in the image of a disembodied soul looking down "from above" on your body carrying out some action. No – at those space-time coordinates, it's the real original *you*, doing whatever happened at the original time. The difference is that your total being is now a combination of all the higher dimensions you now have with the original four dimensions of space and time.

You're not doing something *over again*, not a repeat performance. A "repeat" would imply the same action *at a later time*. This is the original action *at the original time*. It's not just a memory, but the real thing experienced not "again" but originally. Most folks would pick the happiest moment in their lives, you would think. But when you become an advanced being having awareness, cognizance, and control over many additional dimensions, you might see things differently. Anyway, as a multi-dimensional being, you're free to choose whatever coordinates you like, for whatever reasons you like; and can occupy those coordinates as you please, which implies "stay as long as you like."

As a "resurrection-of-the-body" scenario, this is much better than the idea of having all the people who ever lived standing around in one place at some *time* in the future. Imagine the anomaly of seeing exceptional people like the Emperor Constantine, St. Thomas Aquinas, and Shakespeare, all marveling at modern conveniences, like bathrooms.

Accepting our Finite Perception

Sooner or later, it comes down to realizing that God is a whole lot smarter than we are. He is *present* to all space and all times. (People have said that for years without ever thinking about what it means.) Time is not supreme but is just one more of many dimensions. God is also present to all the coordinates of all the extra dimensions, *including* those dimensions that are interchangeable with time. That also includes dimensions totally beyond human imagination.

God created time. Who says he created only *one* temporal dimension? It's a shortcoming of human beings that *we* only *experience* one temporal dimension, along with the three spatial dimensions $\{x, y, z\}$; but we dare not impose that limitation upon God. Doing that would be putting a false god (time) before God. Our proper role is to be deferential.

We have held for centuries that at death "life is changed, not taken away." For the human being to shed his dimensions $\{x, y, z, iCt\}$ and enter into a new set of higher dimensions is entirely possible, if that is what God wants to do. Such a transformation is not to be confused with going into a "parallel universe;" the "parallel universe" would be some *other* set of $\{x, y, z, iCt\}$. This is different. Again, we have no experience of such a thing, no language with which to describe it. All we can go on is faith in God's goodness and greatness.

Moreover, an additional point of humility is that the very word "dimension" may be misleading. In God's total creation, what we humans label a "dimension" may be just a special case of something much more profound, a greater "degree of freedom." The very notion of describing *anything* using terms like "dimension" is just one more example of how limited the human mind is.

From our sensory experience in space and time, we figure out the remarkable similarity between x, y, and z; and from that we understand the concept of "dimension." Next, we invoke mathematics to conclude that there are many dimensions (God has an infinite number of them). As we progress, we must remember that each higher stage we understand is only a little better partial representation of what God has created. It takes physics to figure out that time is a dimension and that we live in the space-time continuum $\{x, y, z, iCt\}$. It is an accomplishment of human thought that we have been able to add *one* more dimension in this way, even though ordinary spoken language is totally inadequate for the purpose. Can we reasonably expect human thought and language to take us much further?

We need to realize that human limitations also impair our *image* of God, but we must never confuse that image with the *reality* of God, which is far beyond our comprehension. Being able to tell the difference makes it easier to live with.

Conclusion

At the start of this chapter, I said I'd present my best guess. All this is extremely speculative! There is no

> **Human limitations also impair our image of God, but we must never confuse that image with the reality of God.**

way any earthly "empirical evidence" can account for anything that is "beyond time." Also, there is no obstacle to anyone else's guessing otherwise. In fact, if this motivates others to think in different directions, then that's a good outcome.

The notion of multidimensionality is made plausible by acknowledging that God, who is almighty, is free to enable all this if He chooses. But I cannot possibly say that God necessarily chooses to "do it my way"! The only safe bet is that God has a plan for our enhanced life beyond time, and that it totally blows away any human concepts about both God and humanity.

The important message, then, is really about what we *don't* know; about what we *cannot* know through human language, experience, and science. We must be extremely patient while God allows us to become better connected with Him. That patience will extend across the transition of physical death. But we have confidence in God that there *is* something far better than we can imagine.

Epilogue

The very first sentence of the Nicene Creed reads, "I believe in God, the Father almighty, Creator of heaven and earth, and of all things visible and invisible." I never really go beyond that sentence. That limited scope is entirely appropriate because my specialization is in science, not in theology.

I have focused attention on the greatly extended meaning of the domain we call "invisible." Science is limited to the domain of the "visible" – that which is accessible to our measuring instruments. There is much more to creation than we can apprehend via sensory perception, and we would like to find unity between those domains. All of that is far more important than most people presume.

In this book, I focus attention on things taken for granted for too long. I have pointed out that God created the very basic symmetry principles that underlie all science. I have emphasized the hidden symmetry between time and space. I have explored new ways of thinking and communicating in order to draw the reader into seeing nature, life, and all reality in a different way.

The dismissive phrase "Science disproves religion" is glibly said by those who don't actually know very much science but who recite what they heard that somebody else said *about* science. Real science demands more concentrated attention. Upon diligent scrutiny, it turns out that faith and science don't actually conflict after all.

Recalling the goal of St. Augustine cited at the outset, I insist that it *is* possible to find the compatibility between the Book of Nature and the Book of Scripture. Few scientists have tried to do so, and too many scientists have deferred the task, instead compartmentalizing the disparate aspects of their lives, thus avoiding conflict. That's a terrible mistake. It is not only demeaning toward God, but it also short-changes our ability to think at a higher level.

The better way to approach seeming incompatibilities is to have confidence that God provides "both/and" answers to questions. And when at last that answer appears, demonstrating the compatibility between religious faith and science, it inevitably displays a brilliance that had previously escaped our attention.

I have drawn concepts and formulations from different fields and combined them to bridge seeming contradictions, to overcome opposing interpretations that had long been treated as mutually impossible. In every case, it turns out that God is a lot smarter than us.

The cover photo of the *Veil Nebula* shows how difficult it is to peer through a cluttered field of debris and serves as a reminder of St. Paul's famous line, "For now we see only a reflection as in a mirror; then we shall see face to face" (1

Cor. 13:12). Indeed, we're looking through a veil; no one pretends to see God clearly.

> **It *is* possible to find the compatibility between the Book of Nature and the Book of Scripture.**

Have I taken a step toward understanding God in a better way? Maybe. My major emphasis has been on pointing out how *little* we *actually* know and how inadequate are our ways of thinking and expressing concepts that pertain to God. Recognizing a barrier is the first step toward overcoming it. I hope that strategy proves valuable. In bumper-sticker length, my message is "step up to a higher level of thinking."

My hope is that you will react by escalating your *own* thinking to a higher plane. If you're motivated to imagine and explore new concepts, then this book has fulfilled its purpose.

Any reader whose grasp of Scripture is better than mine can do that easily. My skill in philosophy peaked with Lonergan's book *Insight,* written in 1956. Others will carry these ideas forward to much higher levels. As exploration in the field of mathematics advances, there will be many new applications ahead (just as physicists devised applications in the early 20th century). *String Theory* is one typical example in physics at present. Expect many other instances ahead.

The advances in the life sciences in the past 2/3 century have been astounding, and as we look into a future involving genetic manipulations, it will bring ever more surprises. We can be certain that the future will look very different from the present.

Going forward, the one reliable guidepost is having confidence that God knows what He is doing and is in control. One theme recurring throughout the Bible is "Be not afraid." We don't need to be afraid of new scientific discoveries because they only unveil, a little at a time, the magnificence of God's creation.

Working together, faith and science can escort us further along a pathway toward understanding God and His creation.

Appendix A

Mathematical Notes

In this appendix, three different mathematical concepts, which pertain to aspects of the created universe, are explained. Humans are capable of understanding these high-school math concepts, and of course we concede that God understands them at least as well as we do.

A – 1: Orthogonal (Perpendicular) Dimensions

The familiar dimensions are left-right, back & forth, up & down. Those 3 form a perpendicular coordinate system $\{x, y, z\}$. Whether outdoors or in an enclosed room, the concept of *perpendicular* dimensions is a familiar one. If we draw a picture of a scene, it will be only a 2-dimensional *projection* of the 3-dimensional *reality*. We're able to mentally adapt and "see" the third dimension because we have experienced all 3 dimensions in reality.

The more general word for that mutually-perpendicular relationship is "orthogonal," of which "perpendicular" is one special case. Each dimension has no overlap with the others. When we advance to think about more dimensions, the property of mutual orthogonality continues. A drawing (a 2-

dimensional projection) of a 4-dimensional object is much harder to grasp because we don't have direct experience of the object in 4-dimensional space. In drawings that represent the 4 dimensions of space-time, usually only x and t are displayed, and the other 2 dimensions are set aside as irrelevant.

The problem of visualization is even harder for more dimensions. However, mathematics works within more dimensions, all of which are *orthogonal* to each other. The rules of algebra, geometry, calculus, etc., apply to any number of dimensions.

A-2: Logarithmic Representation of Time

There are times when data needs to be compressed or expanded to display it usefully. To do that, a common form of display is the *logarithmic scale*. In chemistry and physics, for example, atmospheric pressure drops off as height rises from the earth's surface to outer space, and *pressure* or *density* varies over several orders of magnitude. To display differences in data over many orders of magnitude, a *logarithmic scale* is commonly used.

A wide-screen video shown at a lot of science museums dramatically conveys the extremely wide range of physical sizes by this technique: It starts at the hand of a person sitting on a park bench, and then the camera backs away overhead, taking in the size of the park, the city, the country, the earth, the solar system, and so on out to the view from a cluster of galaxies. Then the camera zooms in, all the way back to the human-size, and then keeps going down to smaller sizes of skin cells, blood cells, atoms, nuclei, quarks, and beyond. That presentation takes place on a logarithmic scale, spanning about 40 orders-of-magnitude.

In everyday life, nothing ever happens where time needs to be displayed on such a compressed or expanded scale. But trying to think about the age of the universe is *not* just "everyday life." For both really long times and really short times, we need to imagine time using a logarithmic scale.

And of course, if *we* can use that method to understand time, so can God. The familiar hymn lyric "A thousand ages in Thy sight are like an evening gone" give us the general idea about *long* time periods, but God does equally well in contemplating very *short* elements of time.

The age of the universe is about 13.8 billion years, which is about 4×10^{17} seconds. Our planet, at 4.5 billion years, is about 1.4×10^{17} seconds old. Reaching back to extremely early times, the tiniest fragment of time that physicists can deal with is around 10^{-43} seconds, a span which carries the name *Planck time*. Less than that, the extremely early universe disappears into the fog of quantum mechanics. Accordingly, the entire *known* time of the universe can be represented on one sheet of paper, using a logarithmic time scale of 60 orders of magnitude. Incidentally, the *size* of the very early universe below about 10^{-34} meters (the *Planck length*) like-wise vanishes into the quantum fog.

There are many different representations, which use a logarithmic time scale for visualization, of the "cosmic calendar" on the internet. One of those is shown below.

If human ingenuity is this good, can we doubt that God can work with 100 or 200 or 10,000 orders of magnitude? In the very earliest times – "long" before the diffuse sea of radiation and quarks, or strings, God knew what was going on. When we represent our knowledge using a logarithmic time scale, whether long time or short time, it doesn't even begin to touch the concept of God's *transcendence*.

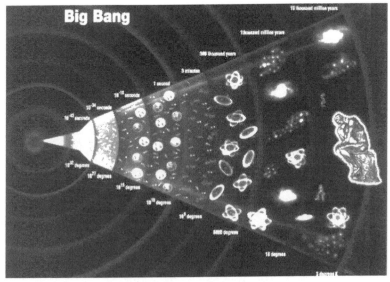

The history of the universe starting with the Big Bang.

A-3 Asymptotic Functions

A function is said to *asymptotically approach* a limit if it never quite reaches a certain value, even though the difference becomes vanishingly small. Consider the function $F(x) = [\,1 - (1/x)\,]$ over the range from $x = 1$ to infinity. When $x = 1$, $F(x) = 0$; when $x = 10$, $F(x) = [1-(1/10)] = 0.9$; when $x = 100$, $F(x) = [1-(1/100)] = 0.99$; and so forth. The value of the function gets closer and closer to 1.0, but never quite reaches 1.0. We call 1.0 the *asymptotic value* of $F(x)$. The adjacent figure is a sketch of such a function.

There are thousands of other such functions that have asymptotic values; most exponential functions do so. For example:

$F(x) = \exp(-x)$ → 0 for very large x; and

$F(x) = [\,1 - \exp(-x)\,]$ → 1 as x goes to infinity.

A counter example is the sines and cosines familiar from trigonometry: no matter how large x becomes, $\sin(x)$ always swings back and forth from 0 to +1 to 0 to -1 to 0 to +1, endlessly.

Asymptotic function

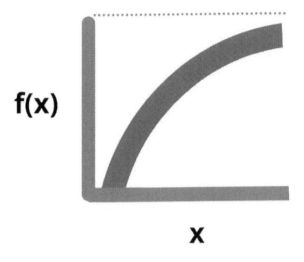

f(x)

x

Appendix B

The STORY of FLATLAND

Flatland, written by Edwin A. Abbott, is a remarkably insightful little book that gently coaxes the reader to realize there is a higher level of reality. Published in 1872, and superficially an enjoyable fantasy about geometry for young math students, the book provides a means of discerning how utterly limited our human perceptions are.

Flatland is about living in a two-dimensional space. The protagonist of the story is named "A. Square," and the book is narrated from his point of view. The first part of the book describes what life is like in *Flatland*, and the second part describes his encounter with a sphere who comes to visit *Flatland* from his higher-dimensional existence.

It is this second part that is so insightful: On the one hand, we chuckle at the confusion and ineptness of Mr. Square because of his inability to grasp the reality of more than two dimensions and his failure to discern that he is dealing with a higher-dimensional being. On the other hand, the underlying message of the book is that we too, in our world, are terribly limited in our ability to grasp higher realities. The sphere, fully comprehensible to us as humans, is a mythological creature to Mr. Square, with awesome and magical powers. The point to be made is that we humans must realize that we ourselves are very limited in our ability to think; therefore, we should be ready to accept the reality of higher-order beings that are beyond the limits of our comprehension, speech, and thought patterns.

Part I:

The narrator (Mr. A. Square) is careful to describe his civilization from a strictly two-dimensional point of view. In fact, when he says that some folks are hexagons or whatever, he explains that the way they tell the difference is by feeling the angles of a person and utilizing their own sense of geo-metrical shapes.[1]

There is no hint of higher-dimensionality because he has no such concepts during the early part of the book. He never gives a "top-down" description of anyone. In particular, the area enclosed by a person is never mentioned because A. Square is not even aware of this concept, having never seen

---

[1] NB: Sometimes feminists complain about the portrayal of women in Flatland as merely one-dimensional lines. But Abbott was writing entertainment for junior-high-school boys. Any junior-high girl who has studied a little geometry can instantly recognize that the flat plane of the men is *perpendicular* to that of women. If men are from XY while women are from YZ, then males in the female flatland will appear as mere lines. All kinds of anatomical humor then becomes possible.

it. He has an awareness of his "insides," perhaps some form of "inner being," but not that it is an area that might be occupied by another plane-geometric figure.

Part II:

Soon thereafter the sphere comes to visit Flatland, and after trying vainly to convince A. Square with words of the reasonableness of his existence, the sphere forcibly takes A. Square into the universe of three dimensions. The experience is unsettling, to say the least:

> An unspeakable horror seized me. There was a darkness; then a dizzy, sickening sensation of sight that was not like seeing; I saw a line that was no line; space that was not space; I was myself, and not myself. When I could find voice, I shrieked aloud in agony, "Either this is madness or it is Hell." ... I looked, and behold, a new world! There stood before me, visibly incorporate, all that I had before inferred, conjectured, dreamed, of perfect circular beauty.
> ...

Upon observing Flatland from a three-dimensional vantage, A. Square experiences the transition from *inference* to true *knowledge*, brought about by his higher-dimensional vision of his own world. His first reaction is to assume the sphere is a god because of what the wise men of Flatland had always taught. But the sphere rebukes him, saying:

> *Then the very pick-pockets and cut-throats of my country are to be worshipped by your wise men as being gods: for there is not one of them that does not see as much as you see now. ... Surely that is no reason why the pick-pocket or cut-throat should be accepted by you as a god.*

The sphere goes on to dispel various other false beliefs of A. Square. Eventually, Mr. Square catches on to the concept of higher dimensions, and tries to lead the sphere to think of 4, 5 or more dimensions; but the sphere can't buy it:

> *...no one has adopted or suggested the theory of a fourth dimension. Therefore, pray have done with this trifling, and let us return to business.*

Nevertheless, Mr. Square persists and pushes harder, until the sphere gets angry and dumps him back into Flatland. Later the sphere comes to agree with him, and so A. Square is emboldened to try to teach higher dimensionality to his family. He struggles in vain to get others to accept the notion of "upward, but not northward." He finds little interest because the people he talks to simply cannot grasp the concepts he is trying to convey. Defiantly he shouts,

"I will endure this and worse, if by any means I may arouse in the interiors of Plane and Solid Humanity a spirit of rebellion against the conceit which would limit our dimensions to two or three or any number short of infinity."

A. Square is variously ignored, considered in need of rest, or judged insane. The message for humans, of course, is that even in our higher-dimensional world, we, too, are blinded by our preconceived notions, and will too readily dismiss as a "nut" anyone who speaks to us from an unfamiliar reference frame.

Special Properties:

The people of *Flatland*, including Mr. A. Square, have certain preconceived notions about what it means to be alive in their two-dimensional space. The sphere freely violates a number of their rules. For example, when attacked by the sharp spears of the palace guard, he merely zooms out of the plane; to the Flatlanders, the sphere appears to have vanished from their sight. But their sight *cannot see out of the plane of Flatland!*

The sphere also enjoys the ability to project himself in many different ways. It is no trouble at all to appear to Flatlanders as a circle of various diameters, changing diameter at will. Moreover, the sphere can project himself into regions of Flatland which are inaccessible to Flatlanders themselves; as for instance when he bumps gently against the inside of A. Square, causing a really weird feeling in A. Square's stomach. And, of course, he can cross walls without benefit of doors. His ability to look down on the plane of Flatland and see it all at once is stunning to Flatlanders, who call this capability *omnividence* and consider it an attribute of a god. Because what passes for "natural" in Flatland is so limited, the sphere is said to have "super-natural" powers.

Going in the other direction is much harder: A. Square must grasp in his mind a higher-dimensional reality for which he has absolutely no reference in experience. He is unable to interact with it under his own control and can only appreciate higher dimensions through great mental effort. Toward the end of the book, his occasionally fading memory of the third dimension shows a weakening or lapse of that mental effort. When the sphere returns in a dream, it reinforces the mental effort needed to sustain A. Square's knowledge of higher dimensions. It's a constant struggle for him to retain his knowledge. In the world of the sphere, A. Square would be said to be limited to only "sub-natural" powers. The pitiful inadequacy of such powers is brought out by the visits to *Lineland* and *Pointland*, where the sphere and A. Square both ridicule the arrogant conceit of the lower-dimensional monarchs.

The Analogy:

Several thousand years ago, Plato tried to tell people that the reality we experience is just a downward projection of a higher reality – he used the analogy of shadows projected on the wall of a cave. That philosophical viewpoint has been widely ignored because it demeans the reality we all experience. Like the high priests in Flatland, we don't like being told that we're not really in control and not even aware of the full extent of our existence. As we struggle within the limitations that we all experience, we would like to believe that there cannot be any other beings who are exempt from our limitations.

The whole point of *Flatland* is to make it easier to get used to such exemptions. By the final pages, hopefully all readers will have gotten the message about expanded realities having higher dimensions. It is possible to strengthen our grasp on our mental acknowledgment of this possibility by reflecting on the types of control we already exert over lower dimensions.

Memory plays a very important role, either for Flatlanders or for us humans. If *we* did not have *memory*, it would be impossible for us to distinguish how freedom of movement around the three spatial directions takes place as time passes. Particular choices of spatial coordinates would be indistinguishable and irrelevant.

Time in Flatland:

When Edwin Abbott wrote *Flatland*, there was not even a glimmer of Einstein's theory that related space to time. Abbott couldn't conceivably have treated time as a dimension. It was simply not within his catalog of concepts, Thus, in *Flatland*, time marches on in the same way it does in normal human experience. The extra dimensions envisioned by Abbott were all spatial dimensions. However, he does mention a square moving perpendicular to itself and thus sweeping out a cube over time.

The Message:

Flatland teaches us not to think that our existence is limited to only a few dimensions. Indeed, it teaches that it is arrogant and conceited to so pretend. Through the entertaining narrative of how a benighted two-dimensional being gradually learns to understand three dimensions, *Flatland* invites us to appreciate our own higher dimensionality, even if we can't grasp it through our ordinary experience. That invitation is an extremely powerful call to reach outward (akin to "upward, not northward"?) toward a higher reality in life, one that is not accessible via the standard means of sensory perception.

Appendix C

Teilhard de Chardin's Upward Step

The French priest Pierre Teilhard de Chardin, S.J., realized the inadequacies of mankind's language and science for describing the complex reality that is the human phenomenon.

To move beyond those constraints, Teilhard did something very similar to the mathematicians who discovered the complex number plane: he introduced a *complex energy plane*, with both *radial* and *tangential* components of energy. Teilhard perceived enough importance in complexity and increasing consciousness that he hypothesized that energy should contain another component, which he named the *radial* component of energy.

Tangential energy is what we've all known for a long time, in the same way that we are familiar with the real number line. The leading characteristic of Teilhard's *radial* energy is that it drives things towards increasing consciousness; it gives the direction to evolution.

This was received about as well as talk of a third dimension in *Flatland*. Teilhard's critics pointed out that there were excellent rules for measuring energy, and instruments to use, by which scientists could reach agreement on their measurements. Statements about tangential energy are subject to falsification, and this is a fundamental principle of the scientific method. Teilhard had no quantitative instruments to offer which could measure his radial component of energy. Consequently, mainstream science paid no attention to this hypothetical new component of energy, preferring to think only of conventional [tangential] energy in exactly the same way that we revert to expressing ordinary (non-complex) multiplication as $3 \times 6 = 18$.

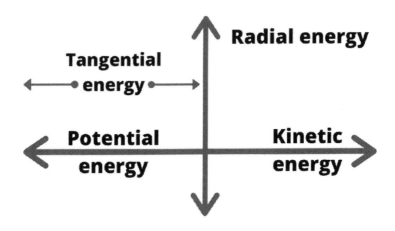

Teilhard also drew a graph showing how consecutive layers of increasingly complex reality blanket the earth. Terms like Lithosphere, Atmosphere, and Biosphere are familiar to us; Teilhard coined the term *Noosphere* to convey the layer of consciousness. The term *noogenesis* refers to the phase of evolution leading into the domain of reflective consciousness. This imagery has proven more congenial to people, and gradually the word *noosphere* became accepted, if not exactly commonplace.

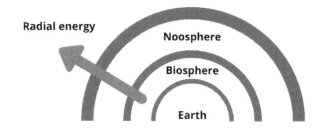

Teilhard brought a new level of thinking to the task of understanding humanity. He began from the premise that science and religion are <u>both</u> correct and could be reconciled. He invented new words where old ones failed. He would not be constrained by standard images. We can learn a lot from his visionary approach.

Appendix D

Fine Tuning: God Making It Work

Was our universe designed and created intelligently, or did it all just happen randomly? Were the basic laws underlying all science just dumb luck, or do they serve some purpose? This appendix examines the exceptional precision of certain numerical ratios, which strongly suggest that the universe owes its existence to an infinitely superior intelligence.

Human beings are able to discern order in nature through the use of reason. Everything science deals with is made of matter and energy. The order we discern in nature is based upon mathematical laws – laws which apply far beyond the realm we can personally experience. The "advance" of science is the discovery of patterns and regularities that display order. To "study" science means to inquire into the organized body of knowledge about order.

For most people, their introduction to physics deals with Newton's laws, energy, momentum, sound, optics, and *maybe* a little atomic physics. With further study, quantum mechanics (QM) opens up an almost mystical wonderland of strange phenomena, which runs all our cell phones.

In basic physics, we learn about certain numerical values, like the temperature at which water boils, the speed of light, the charge on the electron, the acceleration of gravity, and many more. Each of these *physical constants* has some dimensions, or units, like meters per second or miles per hour.

Dimensionless Numbers

The speed of sound (C_s) and the speed of light (C_L) both are measured in the dimensions of meters/second, and so their ratio is a pure number – no matter what units you use, $C_L/C_s = 875,000$. Another key number is the ratio of the mass of the proton divided by the mass of the electron: $M_p/m_e = 1836$. There is a very long list of numbers that can be constructed this way. You might think all these numbers are of no special significance, just a coincidence. For most of them, you would be right.

However, here's what is really interesting: There are certain numbers that are very specially fine-tuned. That is, if their numerical values had been different by even very small percentages, the universe as we know it today could never have come to be. These numbers pertain to the basic forces that govern the universe, the size and time scale of the universe, and the structure of everything. They appear in physics at a very rudimentary level. Coincidences? Actually, they reveal the exquisite *engineering* that went into giving us the universe we live in.

Physicists and chemists have found about 20 such numbers that must be fine-tuned to permit life as we know it. Here, I'll consider just a few of them. In 2000, the British Royal Astronomer Martin Rees wrote *Just Six Numbers*,[1] which devotes one chapter apiece to explaining the significance of each of six particularly special numbers:

1. *Ratio of the Electromagnetic Force to the Gravitational Force:* Gravity is a <u>lot</u> weaker than electromagnetism. The extremely weak force of gravity only builds up to real strength when a <u>lot</u> of mass is present, such as in a planet. In stars and galaxies, there is so much mass that the force of gravity dominates.

The ratio of the electromagnetic force to the gravitational force is about 10^{36} (= 1 with 36 zeroes behind it).

The extreme weakness of gravity is what gives the universe enough time to "get its act together." If gravity were stronger, gases would coalesce into much smaller stars much more quickly, and those stars would burn out more rapidly – in about 10 thousand years instead of 10 billion years. The formation of planets takes longer than that, and the development of life longer still. If gravity were stronger, stars would not last for enough time for us to be here.

---

[1] Martin J. Rees, *Just Six Numbers: The Deep Forces that Shape the Universe* (New York: Basic Books, 2000).

2. *Efficiency of the* Strong *interaction:* Inside a star, hydrogen is "burned" by nuclear fusion into helium. The interactions between protons and neutrons are governed by the *Strong Force.* As hydrogen nuclei combine, a certain amount of mass is converted into energy, according to the famous formula $E = MC^2$. As hydrogen combines to form helium, the amount of mass converted is $E = 0.007$ ($= 7/10$ of one percent). Now, what's interesting is that if this conversion efficiency were $E = .008$, the early protons from the Big Bang would have combined too quickly, and stars would not even have formed. If $E = .006$, protons wouldn't bind to neutrons, and thus there would be no stellar process to produce helium. Either way, no stars as we know them.

Moreover, another process taking place inside stars produces carbon by the combination of three helium nuclei. A chance combination of three things is much too rare to depend on to produce much of anything, but it seems there is a "resonance" – a very precise matching of energy levels – that allows this process to happen, forming carbon from helium. Without the fine-tuning of the strength of the *Strong Force,* that resonance would vanish. Needless to say, without carbon, again no biology as we know it. The value of E has to be quite close to 0.007.

3. *Cosmic Density:* For centuries astronomers have looked at stars, but only in the 20th century was it discovered that the stars cluster together in galaxies. By the late 20th

century, it became clear that the rotating motions within galaxies would cause them to fly apart <u>unless</u> there is a lot of additional unseen mass out there exerting the gravitational pull necessary to hold things together. This unseen mass is termed *Dark Matter*. This idea is fully accepted by astrophysicists because it is based upon a very sound theoretical basis. Our current understanding is that we see very little of the mass in the universe; about 90% of the mass is actually invisible. It is plausible that most of this mass is neutrinos, but other contestants have not been ruled out.

What is important, though, is that the universe has some average density, customarily denoted by ρ. If we take all that can be seen and spread it out uniformly over all space, that density seems to be about $\rho = 0.1$ atoms/m^3; and if we add interstellar dust, it becomes $\rho = 0.2$ atoms/m^3. The presence of dark matter runs it up to around $\rho = 2$ atoms/m^3. Those are the current values.

But the very early universe was different. Starting off from the Big Bang, the expansion of the universe can be traced in space and time. If the density exceeded $\rho_c \sim 5$ atoms/m^3, the strength of gravitational attraction would be so great as to pull everything back together again in a giant collapse. ρ_c is called the "critical density." The ratio of the *actual* density to the *critical* density is denoted by Ω and is another of the six special numbers. If the *actual* density were slightly lower than the *critical* density ($\Omega < 1$), expansion would proceed rapidly, and density would become lower still; in that case, stars and galaxies would never form, and the universe would simply fly apart.

The range of permissible values of Ω is very narrow – and sure enough, we're in that range. The actual expansion rate of the universe is an observable, measurable quantity, known as the *Hubble Constant*. We observe that, in over 10 billion years, Ω has stayed remarkably close to one. For that to be true today, it must have been the case that at one second after the Big Bang, $\Omega = 1$ had to hold within one part in 10^{15}. Quite possibly, Ω has to be *exactly* one, for reasons not yet discovered.

4.   *Smoothness and Ripples:* Any theory of cosmology must match the observations from astronomy. One problem is that the observable universe is certainly non-uniform. If one imagines a "big bang" followed by expansion, at first it would be plausible to suppose that the expansion proceeded uniformly; in which case, there would be no particular reason for stars and galaxies to coalesce.

The primary evidence that a Big Bang <u>did</u> occur is the *Cosmic Background Radiation,* which shows that the universe is filled with radiation corresponding to a temperature of 2.7 °K. That radiation, leftover from a very hot big bang almost 14 billion years ago, was first discovered in 1963, and seemed to be coming uniformly from all directions. However, in recent years, the NASA *Cosmic Background Experiment* (COBE) showed that there were small fluctuations ("ripples") in this radiation, and those are enough to trigger the formation of galaxies: denser regions led to galaxies, and sparser regions led to voids. The magnitude of the initial fluctuations is very small: about one part in 10^5 (1/100,000). But as stars and galaxies coalesce, slight density fluctuations are magnified over time.

Separately, the gravitational "binding energy" of a galaxy, divided by the energy of its rest mass ($E = MC^2$) is denoted by Q, and is about 10^{-5}. This number Q provides an estimate of the size of "ripples" in the density of space. The fact that it comes out about the same as the "ripples" in cosmic background radiation confirms the relation between the two – the primordial fluctuations of radiation density presumably led to the observed density variations across intergalactic space.

Since $Q \sim 10^{-5}$, it means that gravity is weak in a galaxy (or even in a cluster of galaxies), so Newton's laws are applicable. That in turn allows machine computations to be done on an expanding universe subject to its own gravity, and the results simulate how gravitation leads to clusters of galaxies. The numerical outcome is statistical, of course, but doesn't conflict with what is observed.

However, if Q were significantly smaller than 10^{-5}, galaxies would coalesce much slower and looser, star formation would be much slower, and the heavy elements formed in a supernova would easily go away so that planets could not condense around stars.

If Q were significantly larger than 10^{-5} (large ripples), very large galaxies would coalesce quickly and collapse into black holes. Stars (if any) would be so close-packed that they could not have planets. Either way, Q has to be close to 10^{-5} or else no planets form, and once again, we're not here.

5. *The Cosmological Constant,* Λ: When Einstein first proposed the general theory of relativity, he introduced a term (Λ) known as the *Cosmological Constant*, whose role was to provide a balancing force that opposed gravity and kept everything from collapsing. It was to represent a force even weaker than gravity, one that would only have effects on a galactic scale, undetectable on our planetary scale. This force created a "cosmic repulsion." However, this factor Λ didn't enhance the beauty of Einstein's equations. It became unnecessary several years later when the universe was observed to be expanding. So, theorists set $\Lambda = 0$, and put it aside. Years later, Einstein called it his "greatest blunder."

Many decades later, Λ is making a comeback. There are irregularities in the observed primordial background radiation. Astronomical observations of red-shifts from distant supernovae tell how fast galaxies are receding and hence give a measure of the expansion rate. But still Λ must be a very small number. If $\Lambda = 0$, then the expansion of the universe is decelerating. On the other hand, with a finite Λ, there would be an anti-gravity effect that actually accelerates the expansion. Thus, the fine-tuning of Λ affects the predictions of the long-term future of the universe. Today's best guess is "stay tuned."

6. *Dimensionality, D:* The sixth number that Rees features is the dimensionality of space, denoted by D. This equals 3, and thanks to the theory of relativity, the space-time continuum is taken as 4-dimensional. The connection of *time* to the 3 *space* dimensions is easy to do mathematically by associating an imaginary number with time; the equations of physics take on a symmetry and beauty when this is done.

Nevertheless, it is easy to show that a universe's having 2 or 4 (or more) spatial dimensions doesn't work. The electromagnetic and gravitational forces only fall off as $1/r^2$ in a system having 3 spatial dimensions. That means that atoms wouldn't form, and the orbits of planets would be unstable if there were other than 3 spatial dimensions. In fact, William Paley, a theologian-scientist circa 1800 first stressed the importance of this dimensional requirement. Again, the significance is that we wouldn't be here unless $D = 3$.

What Does It All Mean?

So, what are we to make of these remarkably fine-tuned numbers? Are they all just coincidences, even the one (Ω) that must be precise to the 15th decimal place? Most people call this very clear evidence for design – the numbers seem to point out clearly that the Creator of the universe had perfect control. The precision of the numbers is awe-inspiring to mankind.

Unfortunately, the phrase "intelligent design" has been blurred by confusion and politics and is often associated with biological creationism. Recognizing this distortion, we need a new phrase to convey the idea that God's comprehension and creativity runs all the way from mathematics to human beings.

The message that stands out from the exceptional precision of the dimensionless ratios is that our universe owes its existence to an intelligence far superior to our own, Who wanted things to come out in a very special way *and* wanted us to be here eventually. But the "eventually" that involves us is not the end of the road. The *really* interesting question is: "What's next?"

About the Author

G. Galilei and T. Sheahen in the *Villa Borghese*, Rome, Italy

Thomas P. Sheahen earned BS and PhD degrees in physics from the Massachusetts Institute of Technology. During his 45-year career as a research physicist, predominantly in energy sciences, he worked for Bell Telephone Laboratories, the National Bureau of Standards, various research corporations, the U.S. Department of Energy, Argonne National Laboratory, and the National Renewable Energy Laboratory. He was chosen as a Congressional Research Fellow by the American Physical Society, dealing with energy-related national legislation. Dr. Sheahen wrote the textbook *Introduction to High-Temperature Superconductivity.*

Dr. Sheahen, a lifelong Catholic, is director emeritus of the Institute for the Theological Encounter with Science and Technology (ITEST), which focuses on demonstrating the compatibility of faith and science as paths toward knowledge.

About the Institute for Theological Encounter with Science and Technology (ITEST)

The Institute for Theological Encounter with Science and Technology (ITEST) is an association of theologians, scientists and others committed to a Catholic worldview in which faith and science collaborate in exploring the truth. ITEST explores truth theologically in the wisdom traditions of the human community and in the data studied in the sciences. ITEST fosters and disseminates the Catholic position that science and faith in God are complementary paths to human fulfillment.

For access to ITEST's webinars, bulletins, and other media, visit https://www.faithscience.org and become a member today!

Made in the USA
Columbia, SC
29 November 2021

50055255R00130